# 走进EXPO2010
STEP INTO EXPO 2010
## ——2008七校联合毕业设计作品
SEVEN-UNIVERSITY UNITED ARCHITECTURAL GRADUATION PROJECTS

黄一如 许懋彦 龚 恺 徐苏斌 卢 峰 王 竹 吕品晶 编

中国建筑工业出版社

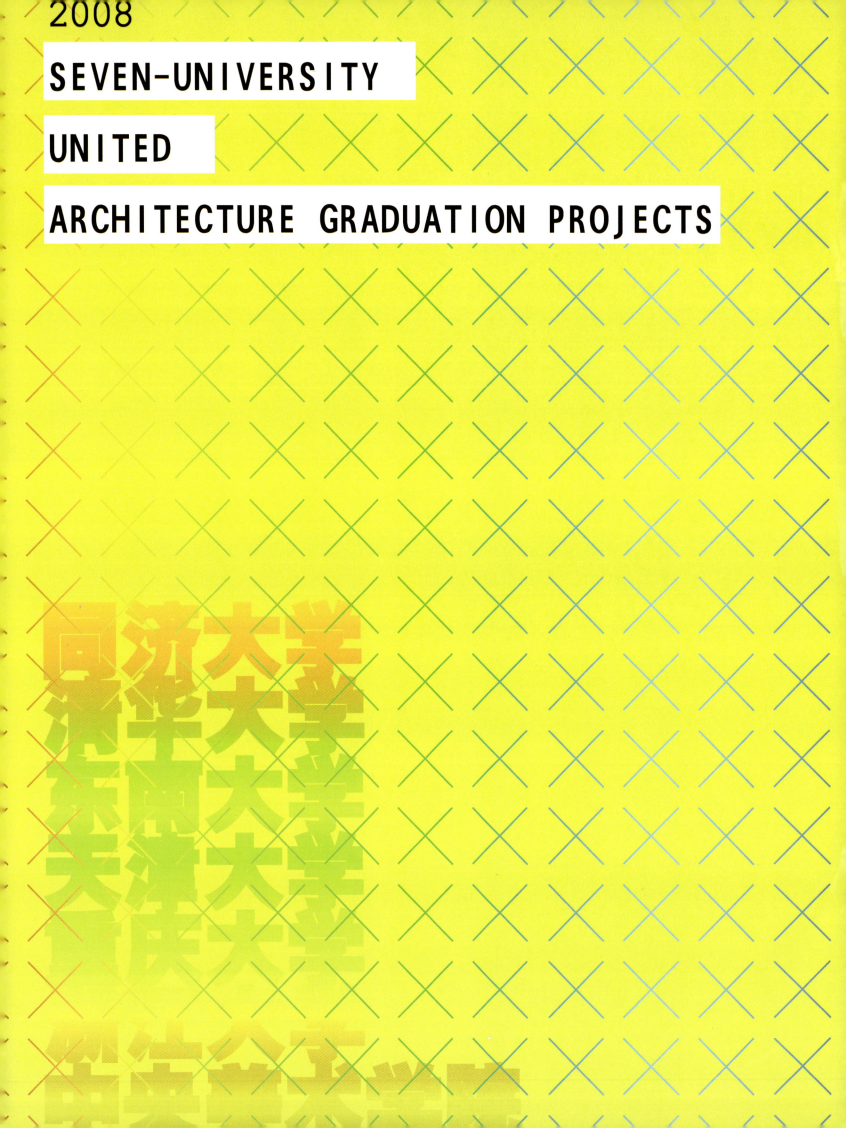

# 本书编委会

同济大学：吴志强　吴长福　黄一如　袁烽　李翔宁

清华大学：许懋彦　刘念雄　张利　卢向东

东南大学：王建国　仲德崑　龚恺

天津大学：张玉坤　青木信夫　徐苏斌　谭立峰　陈天　贾正阳

重庆大学：卢峰　顾红男

浙江大学：王竹　朱宇恒

中央美术学院：

吕品晶　刘彤昊　周宇舫　王铁　傅祎　吴晓敏

# 参与联合设计学生名单

## 同济大学

胥一波　花扬　许嘉　靳阳洋　王岱琳　张一戈　朱哲人　沈文昊　陈日川　孟媛　蔡青　余国璞　田唯佳　王瑜

## 清华大学

曹雩　何珊　郑秀姬　沈华　乔拓　梁其伟　李娜　郑瑄　戴南

## 东南大学

唐颖立　刘彦珉　卢犁梦　李瑞鹏　张志兵　胡晨希　柴金戈　肖冰　郭东海　黄轩　申苏蓉　邱珏　崔赟　刘默琦　邱柏霖　姚昕悦　江天　何晋　袁懿　李邦健

## 天津大学

马睿　唐涵　王特立　李程　任桂园　廖胤希　张晓未　惠超微　何美　孙洋

## 重庆大学

万博　伍鹏晗　张琴　付烨　王文婷　周雪峰　孙海霆　崔璨　王怡斐　韩文晶

## 浙江大学

刘伯宇　邓超　孙翌　祝马丽　刘蕾　张黎源　张安琪　杜鹃　孙佩雯　陶俊　公贤彦

## 中央美术学院

曹卿　杨剑雷　张玉婷　袁野　段敏　马超　苏迪

# 目录

序　2

2008七校联合毕业设计课程内容及安排　4

"七校联合设计"大事记　8

## 同济大学

尺度蒙太奇　12
设计一组：胥一波　花扬　许嘉　靳阳洋　王岱琳

城市印章　22
设计二组：张一戈　朱哲人　沈文昊　陈日川　孟媛

i-industry 城市设计　32
设计三组：蔡青　余国璞　田唯佳　王瑜

## 清华大学

城市经纬　44
设计一组：曹雩　何珊　郑秀姬　沈华　乔拓

市·博　54
设计二组：梁其伟　李娜　郑瑄　戴南

## 东南大学

旧工厂适应性再生　66
设计一组：唐颖立　刘彦珉　卢犁梦　李瑞鹏　张志兵

江南造船厂历史地段之肌理再生　76
设计二组：胡晨希　柴金戈　肖冰　郭东海　黄轩

从历史禁区到城市开放空间　86
设计三组：申苏蓉　邱珏　崔赟　刘默琦　邱柏霖

风场·水场·工场　96
设计四组：姚昕悦　江天　何晋　袁懿　李邦健

## 天津大学

向迪士尼学习　108
设计一组：马睿　唐涵　王特立　李程　任桂园

破土：超越时间的新生　118
设计二组：廖胤希　张晓未　惠超微　何美　孙洋

## 重庆大学

绿色城市　130
设计一组：万博　伍鹏晗　张琴　付烨　王文婷

双城计　140
设计二组：周雪峰　孙海霆　崔璨　王怡斐　韩文晶

## 浙江大学

时之容器　152
设计一组：刘伯宇　邓超　孙翌　祝马丽　刘蕾

工业·进行时　164
设计二组：张黎源　张安琪　杜鹃　孙佩雯　陶俊　公贤彦

## 中央美术学院

生活让城市更美好　176
设计一组：曹卿　杨剑雷　张玉婷　袁野　段敏　马超　苏迪

后记　188

# Contents

2 Prologue

4 Theme

8 Events

## Tongji University

12 Montage on Scale
*Team 1: Xu Yibo,Hua Yang,Xu Jia,Jin Yangyang,Wang Dailin*

22 City Seal
*Team 2: Zhang Yige,Zhu Zheren,Shen Minghao,Chen Richuan,Meng Yuan*

32 i-industry Urban Design
*Team 3: Cai Qing,Yu Guopu,Tian Weijia,Wang Yu*

## Tsinghua University

44 Intertexture
*Team 1: Cao yu,He Shan,Zhen Xuji,Shen Hua,Qiao Tuo*

54 Commershow
*Team 2: Liang Qiwei,Li Na,Zhen Xuan,Dai Nan*

## Southeast University

66 Adaptive Revitalization of Old Factory
*Team 1: Tang Yingli,Liu Yanmin,Lu Limeng,Li Ruipeng,Zhang Zhibing*

76 Urban Fabric Transplant
*Team 2: Hu Chenxi,Cai Jinge,Xiao Bing,Guo Donghai,Huang Xuan*

86 From Forbidden Zone to Public Space
*Team 3: Shen Surong,Qiu jue,Cui Yun,Liu Moqi,Qiu Bolin*

96 Wind Water Work Field
*Team 4: Yao Xinyue,Jiang Tian,He jin,Yuan Yi,Li Bangjian*

## Tianjin University

108 Learn from Disneyland
*Team 1: Ma Rui,Tang Han,Wang Teli,Li Cheng,Ren Guiyuan*

118 Breaking Ground: The Rebirth Transcended Time
*Team 2: Liao Yinxi,Zhang Xiaowei,Hui Chaowei,He Mei,Sun Yang*

## Chongqing University

130 Green City,Green Net
*Team 1: Wan Bo,Wu Penghan,Zhang Qin,Fu Ye,Wang Wenting*

140 Two Story Towns
*Team 2: Zhou Xuefeng,Sun Haiting,Cui Can,Wang Yifei,Han Wenjing*

## Zhejiang University

152 Time Container
*Team 1: Liu Boyu,Deng Chao,Sun Yi,Zhu Mali,Liu Lei*

164 Better Industry,Better Life
*Team 2: Zhang Liyuan,Zhang Anqi,Du juan,Sun Peiwen,Tao Jun,Gong Xianyan*

## China Central Academy of Fine Arts

176 Better Life Better City
*Team: Cao Qing,Yang Jianlei,Zhang Yuting,Yuan Ye,Duan min,Ma Chao,Su Di*

188 Postscript

# 序

每年两次的国际建协(UIA)建筑教育委员会会议上，总无法回避的话题是世界建筑教育未来方向。有RIBA的英式和ACSA的美式传统坚持，有墨西哥的南美模式崛起，更有阿拉伯建筑教育的抗争。东亚只有我与另一位日本委员在其中，总也积极地发出自己的论点。我们没有欧美的建筑教育的历史传统，几乎第一代师承的都是欧美教育，我们也没有发展中国家对于建筑教育在全球化下的怒气，我一直在思考，中国建筑教育未来究竟走向何方？

有几点背景，可能冥冥之中我们不能忘记：

首先，中华古代文明中的建筑智慧，是我们自觉地不应该、实际上也是集体无意识中根本无法忘记的精神财富。老子的建筑观在乎于实与虚间、礼乐的建筑观在乎于社会秩序之中，识天地而后建构并融化其内，为我中华一脉相承之建筑追求，此为我见到的永恒力量。这种智慧的力量，超越了时代政治、超越经济武力。一种静静地流淌，却又至高无上的智慧，其力让人理解后折服。我处在四川大地震中的都江堰时本能地想跪在这种智慧面前。

其次，中国老一辈的建筑教育家们奠定的体系，是我们讨论时下和未来建筑教育的真实基础，任何脱离了现实的讨论，可能都是无知或者幼稚的。有人可以把中国的建筑教育分成几大校，或者称为几大Schools。而我读了前辈们早期的记录后，看到的是一批批热血的学子归来，以其留学的经验，游走于那苦难的祖国南北，在那些还能生存的地方创办一所所建筑院校。大地在战乱，老师刚学成。如果能设身处地，除了敬意，我还想说那是功德。

其三，改革开放的建设机遇与思想环境，不仅有北京奥运、上海世博、天津滨海及广州亚运的大事件可让学生参与其中，实际上，眼下的所有城镇乡村，年年都总把新桃换旧符，此实为举世难得。三十年间从描图员到配施工图，从同台竞争到教育输出，全国已经上下练就了一批优秀的建筑师、教育家。这是必须认识的知识财富，否则集体学费付得太高了。

基于以上三点背景，我想表达的是：在国际竞争的背景下，国内学校间的竞争已经不重要了，中国的建筑教育需要形成一个大格局；共同的历史文化背景上，中国的建筑教育需要形成一个大理想。举国一盘棋已经让我们享受了北京奥运会的喜悦，中国的建筑教育又何尝不能这样？

大理想在于新中华文明创造中的建筑教育的贡献；大格局在于分工协作中形成立足于今日世界建筑教育中的中国范式。

上次的六校联合设计和这次的七校联合设计的成功举办，均为走向大理想和大格局的历史性范例，所有的学生、教师乃至学校都从思想交流与碰撞中获益。这也说明，一旦打破了院校间的藩篱，中国的建筑教育是完全能够搭建起一个共享平台的。

这次作为东道主，要办好联合设计，有个好的题目尤为重要，既具体、抓得住，又要让学生有发挥的余地和空间。作为2010上海世博总规划师，我在确定设计课题的时候，首先就想到了江南造船厂改造再利用，因为世博会本身就是个开放的主题，而当时江南造船厂地块的用途更是尚无定论，只有世博会后用作近现代工业博览群的初步设想。这样的课题有着很大的弹性，还有要保留的超大型船厂厂房等现实的依凭，既有城市设计的内容，又有建筑单体设计的要求。这种"真题假做"的好处还在于，世博会举办之时或举办之后，学生们能够通过比较自己的设计构思与实际操作之间的异同，通过反思加深对设计的认识。

这次参加联合设计的既有传统意义上的老校，同济、清华、东南和天大，也有重庆大学、浙江大学、中央美院这样迅速崛起的后起之秀，具有广泛的代表性。而设计过程中的开题、中期汇报和最终讲评分别在同济、东南和浙大举行，加上各校教师针对设计课题所作的精彩讲座，增加了七校师生的接触面和交流机会，为取得出色的设计成果奠定了扎实的基础。

就城市设计而言，本来就没有固定的模式。在这次联合设计中，除了均注重与城市的关系以外，各校体现出了各自不同的教学理念和特点：有的更加重视空间形态，有的侧重于设计的深度，有的则把重点放在导则的制定和控制；有的着力于世博会期间的组织与利用，有的则注重世博会后的开发和利用……使得最终的设计成果纷繁多样，具有启发意义。

现在，各个建筑院校都在积极推动国际交流，相形之下，国内高校间的交流反而不够充分。我很同意一些学校老师提出的观点，如果今后的联合设计能够更进一步，在过程中打破学校的界限进行分组，则可以进一步增加各校师生间的接触和交流机会。

这是一本沉甸甸的设计作品集，基本上代表了我国建筑教育在本科阶段的状况和水平，各校学生对设计主题的精彩演绎和创见，对2010上海世博会的成功举办也有积极的参考价值。

谨以这段文字感谢各校师生的共同努力，祝愿国内院校间的联合。我在这些作品中见到了中国未来建筑教育的方向，看到了新中华文明在形成中蓄力，世界的建筑教育界会注意今天我们在这里所作的努力。

2008年夏
于西班牙Zaragoza

# Prologue

At the meeting of UIA Architectural Education Committee organized twice on yearly basis, the inevitable subject always rests with the direction of future development of worldwide architectural education. There may be seen the adherence to British style advocated by RIBA and American style upheld by ACSA, the rise of South America mode from Mexico, and the struggle from Arabian architectural education. From East Asia, only I and another committee member from Japan are invited. We also actively make our viewpoints heard. Our architectural education bears no possession of the historical tradition of architectural education of Europe and America, and basically it relies on the education from Europe and America since its emergence. We also carry no anger with the architectural education like some other developing countries under the globalization. I just keep wondering where the Chinese architectural education will march forward in the future.

Several pieces of background information may not escape us within our deep consciousness.

Firstly, the architectural wisdom of Chinese ancient civilization is the spiritual treasure that we, with self consciousness, should not forget, that actually we are unable to forget thanks to our collective unconsciousness. The architectural concept of Laozi stays between entity and empty; the architectural concept of rites and music is presented in the social order. The architectural pursuit developed with China is to cognize the heaven and earth before construction and then melt the cognition into the construction itself. I have seen an eternal power from this concept. This power of wisdom has surpassed politics, economy and force. This wisdom flows tranquilly and peacefully but stands that supreme after all and people are powerfully convinced by its power upon understanding. I knelt down before this wisdom instinctively in the face of Dujiangyan during Sichuan earthquake.

Secondly, the system founded by the former generations of Chinese architectural educators is acting as a genuine basis for our discussion on current and future architectural education. Any discussion divorced from reality would be empty or naïve. Some may divide Chinese architectural education into several dominant schools. However, what is unfolded before my eyes after reading the early record made by our predecessors is that groups of ardent students returned to China after their oversea study, traveled every corner of our tribulation-filled motherland, and built up architectural institutions one by one on those survival-may-exist pieces of land with their experience. The land was marked with chaos caused by war; teachers were just equipped with the fresh knowledge. Supposedly under those circumstances, beside my respect, I would like to emphasize that it is the virtue that matters.

Thirdly, we are granted the architectural opportunities and thought environment brought in by the reform and opening-up. Not only students may participate in national events like Olympic Games in Beijing, World Expo in Shanghai, Binhai construction in Tianjin and Asian Games in Guangzhou, but they are facing an excellent moment that currently every town, every city and every village wears a new look every year. Throughout 30 years, from tracing workers to specialists capable of preparing construction drawing, from the competition on the same platform to education export, China has witnessed the emergence of a substantial amount of outstanding architects and educators. This is the knowledge wealth we must be aware of; otherwise the collective tuition would be too high.

On the basis of the above-mentioned three aspects, I would like to express: with the background of international competition, the competition between domestic institutions has become not that significant. The architectural education of China requires the formation of a grand pattern. Upon the common historical and cultural background, the architectural education of China is in need of the formation of a grand ideal. Engaging in one game with united spirits across the whole country has rendered the joy of succeeding in organizing Beijing Olympic Games, so why the architectural education of China cannot adopt this approach?

The grand ideal consists in the contribution made by the architectural education throughout the creation of new Chinese civilization; while the grand pattern lies in the Chinese pattern established out of distribution and collaboration as well as among present architectural educations all over the world.

Historical examples have been created in the direction of grand ideal and grand pattern through the successful organization of Joint Studio uniting 6 schools earlier and Joint Studio uniting 7 schools currently. All the students, teachers and schools benefit from thought exchange and collision between viewpoints. This demonstrates that once the fences between schools are broken, the architectural education of China is fully capable of building up a sharing platform.

As a host of this United Design, we consider it particularly significant to define a good subject, which should be specific, comprehensible and able to provide the space for students' acts. As chief planner of Shanghai World Expo in 2010, the first subject jumping into my mind when defining the design subject is the reconstruction and reutilization of Jiangnan Shipyard. The World Expo itself is an open theme. The usage of the land where Jiangnan Shipyard is located had not been nailed down at the moment except that the preliminary assumption existed that this piece of land may be employed for the exhibition of industry cluster of modern times after the World Expo, therefore, the subject bears great flexibility, requiring the maintenance of super giant workshops of the shipyard, the content of urban design and the request of architecture signal design. Furthermore, another advantage of this "supposed design based on actual existence" subject lies in its allowing students to deepen their cognition of design through comparing the difference between their design and practical operation and through reflection in the process of World Expo or after the World Expo.

Participators in this United Design are "old" universities in the traditional sense like Tongji University, Tsinghua University, Southeast University and Tianjin University, as well as rising stars like Chongqing University, Zhejiang University and China Central Academy of Fine Arts. They are representatives. Tongji University, Southeast University and Zhejiang University host respectively the introduction of the subject, the mid-term presentation and the final review during the design process. In addition, teachers from various universities deliver wonderful lectures with focus on the design subject; the contact and exchange increase between teachers and students from 7 universities. All these efforts create a solid foundation for achieving distinguished design results.

Regarding the urban design, it embraces no fixed pattern. In this United Design, besides participators all pay attention to the relationship with cities, universities present their own education concept and features respectively. Some universities may tend to stress spatial form, some may emphasize the depth of design, some may center on the establishment of guidelines and the control, some may devote to the organization and utilization during the World Expo, some may concentrate on the after-Expo development and utilization... leading to colorful and inspirational design fruits.

At present, all architectural schools are promoting international exchange actively. In contrast, the exchange between domestic schools becomes inadequate. I agree on the opinion put forward by some teachers: if the joint studio divides schools into groups by breaking up the boundary of schools next time, the contact and exchange between teachers and students from various schools may increase further.

This is a heavy design work. It basically represents the status and level of the undergraduate education of Chinese architectural education. The outstanding interpretation and creative presentation of the design subject will contribute the positive reference value to the successful organization of Shanghai World Expo in 2010.

I would like to avail myself of this letter to express thanks to hand-in-hand efforts of teachers and students from all schools. I wish domestic schools to be united. I have seen the direction of future development of Chinese architectural education from these works and the effort of harboring power during the formation of new Chinese civilization. The worldwide architectural education will notice the efforts we make at this moment right here.

Wu Zhiqiang
In summer, 2008
In Zaragoza, Spain

# 2008七校联合毕业设计课程内容及安排
同济大学、清华大学、东南大学、天津大学、重庆大学、浙江大学、中央美院

题目：城市建筑的更新与再生——2010年上海世博会中国现代工业博物馆群设计
时间：课程分四个阶段
3月5日　　　　　现场测绘、调研、相关讲座（上海集中），同济大学
4月　　　　　　各校自行安排讲课、收集案例、策划整理、理念构思
4月26日至27日 中期集体交流、评图（南京集中），东南大学
5月至6月中　　深入设计
6月7日至8日　 最终评图，浙江大学

## 一、课题内容及目标
（一）选题意义
　　现代的城市运动不断地扩展着城市的版图，工业被迫渐渐地向城市更边缘发展，工业基地搬迁的现象纷纷出现，对原有工业片区的大型工业建筑群的更新与再生成为了当今世界城市规划专业和建筑设计专业人士关注的重大课题之一。
　　江南造船厂的前身是140年前清政府直属的江南制造总局。1865年清朝洋务运动早期，两江总督李鸿章买进虹口的美商旗记铁厂，开始自行制造火药和枪支。1867年，江南制造局从虹口租界迁至黄浦江畔的高昌庙现址，逐渐成为赫赫有名的江南造船厂。
　　江南造船厂，是"中国近代民族工业的发源地"。这里不仅诞生了中国第一家近代企业，也诞生了中国第一批正规产业工人、第一艘机动兵轮、第一磅无烟火药、第一门工业火炮，甚至第一炉钢。在清代寻求自强之路的洋务运动背景下产生的江南造船厂，不仅是中国百年近代工业的缩影，同时也是中国近代史的缩影，几乎见证了中国近代以来的所有重大事件。
　　根据2010年上海世界博览会的总体规划，江南造船厂将被整体搬迁到长兴岛。对现状江南造船厂地区进行城市空间进行分析研究和改造，提出改善和整治优化方案，并设计2010年上海世博会中国现代工业博物馆群，对延续该地区的历史文化，加强该地区的城市活力，具有重要的意义。
　　（二）基本目标
　　本项目包括城市设计与建筑设计两部分内容。学生应该从整体区域研究出发，通过现场调研、问题分析、项目策划，做到单体设计（改造与新建）深度，并对建筑细部节点构造有一定的了解。
　　1.以2010年世博会为契机，研究该地区城市空间的历史、发展和变化，认真探讨对该地区改造利用的可能性与方式、方法。
　　2.在研究的基础上提出改善城市空间的方案，提高该地区的活力。
　　3.进行建筑策划，对该地区建设项目提出具有可行性的设想。
　　4.探讨如何保留改造具有表征性、地标性的建筑物、构筑物，强化历史文脉和再生利用的价值与意义。
　　5.初步认识了解旧建筑改造与利用的技术手段和构造特点。

## 二、各阶段的重点及要求
（一）预备阶段：
2008年1月21日，由同济大学、清华大学、东南大学、天津大学、重庆大学、浙江大学、中央美院共同讨论由同济大学提出的设计题目，并参观地段。
同济大学拟定任务书及提供课题基础资料及联系相关专家，并负责任务书中的总图及部分现状建筑测绘资料整理（CAD）。
（二）第一阶段：现场调研
了解掌握江南造船厂地区的基本概况。
相关讲座与参观：
讲座1：世博会总体规划，吴志强教授、同济大学建筑与城市规划学院院长
讲座2：世博会景观塔与未来馆设计，同济大学章明副教授
讲座3：上海啤酒厂历史建筑改造设计，同济大学黄一如教授
讲座4：同济大学大礼堂改造，同济大学袁烽副教授
参观上海雕塑艺术博物馆，莫干山路M50创意产业聚集区
现场调研需进行实地踏勘，并做必要的现状测绘完善工作。七校学生共同确认城市设计外部条件，同济大学、东

南大学和浙江大学学生负责补充完善保留建筑的CAD测绘图纸。要求将可能保留的建筑物或构筑物制作出比例为1:200的平、立、剖测绘图。

（三）第二阶段：方案构思

在测绘的基础上，启动江南造船厂地区的城市空间改善与建筑改造设计构思，学生可以进行项目策划，确定设计任务主题（2～4人一组）。

成果要求：各学校组织两次汇报讨论会，第一次是案例讨论读书报告会，要求每人介绍2个案例，准备10分钟PPT格式的汇报。第二次是调研成果、项目策划及初步构思成果汇报。要求每组准备15分钟PPT格式的汇报。

（四）第三阶段：中期评图

地点：东南大学

时间：4月26～27日

讲座安排：

讲座1：后工业时代产业类历史地段和建筑的保护再生，王建国，东南大学教授

讲座2：城市空间的纪念性与人民性，张利、清华大学教授

安排参观大屠杀纪念馆和一处工业建筑改造

学生汇报要求：灵活运用地段、平、立、剖、节点详图、工作模型、三维表现图等充分表现阶段设计思想、方法和结果；具体方式自定。

（五）第四阶段：深入设计

成果要求：

1.各校完成一个城市设计方案及模型（1：1000）

2.每人完成6～8张A1排版图纸，包含有基本平、立、剖面图，设计过程的分析、说明；表现效果图及模型照片等；一个1:300的工作模型；15分钟的PPT汇报文件。

总评图：

地点：浙江大学

时间：6月7～8日

讲座安排：

讲座1：生态与景观视野下的杭州西溪湿地保护，王竹教授、浙江大学建工学院副院长、建筑系系主任

讲座2：杭州西湖综合保护工程，周为，杭州市园林设计院院长、总工程师

考察浙大具有历史风貌与自然风光的老校区（之江、玉泉、华家池）以及西溪国家湿地公园

三、项目指示书

该指示书仅供参考，学生可以依据自己的调研策划与教师讨论进行调整。此项工作可以分组进行，探讨各种可能性。

（一）规模与性质

1.上海江南造船厂地区改造项目规划设计的用地规模为23.4ha，用地范围见附图1。

2.项目容积率控制在1.2以下，用地范围内绿化率要求达到30%，建筑密度在45%以内,停车数要求按国家相关规范。

3.项目内容：中国现代工业博物馆群，包括世博会历史展览馆总馆，中国近现代工业博物馆总馆，中国航天、航空博物馆，中国航海造船博物馆，中国现代重工业博物馆，中国现代轻工业博物馆等。

（二）调研重点

1.了解江南造船厂地区的发展变化历史。

2.了解该地区现存潜力和城市问题。

3.实际调研该地区的现状城市生活形态及使用者的要求与构想。

4.了解掌握建筑改造中的建筑技术规范，对新技术的应用及相关建筑与结构技术发展的动态。

5.了解建筑智能、节能及生态技术应用与建筑改造结合的可能性。

（三）设计构思理念的定位

1.改善江南造船厂地区的城市环境质量，增进城市活力。

2.改善该地区与周边地区的联系，增加配套服务设施（停车场、商业服务、公厕、街道家具等等）。

3.针对重点人群的活动设施。

4.根据工厂原有设施和设备，可以考虑工业遗产保护展示内容。

5.探讨节能、生态型试验建筑的可能性。

6.满足残疾人建筑设计规范。

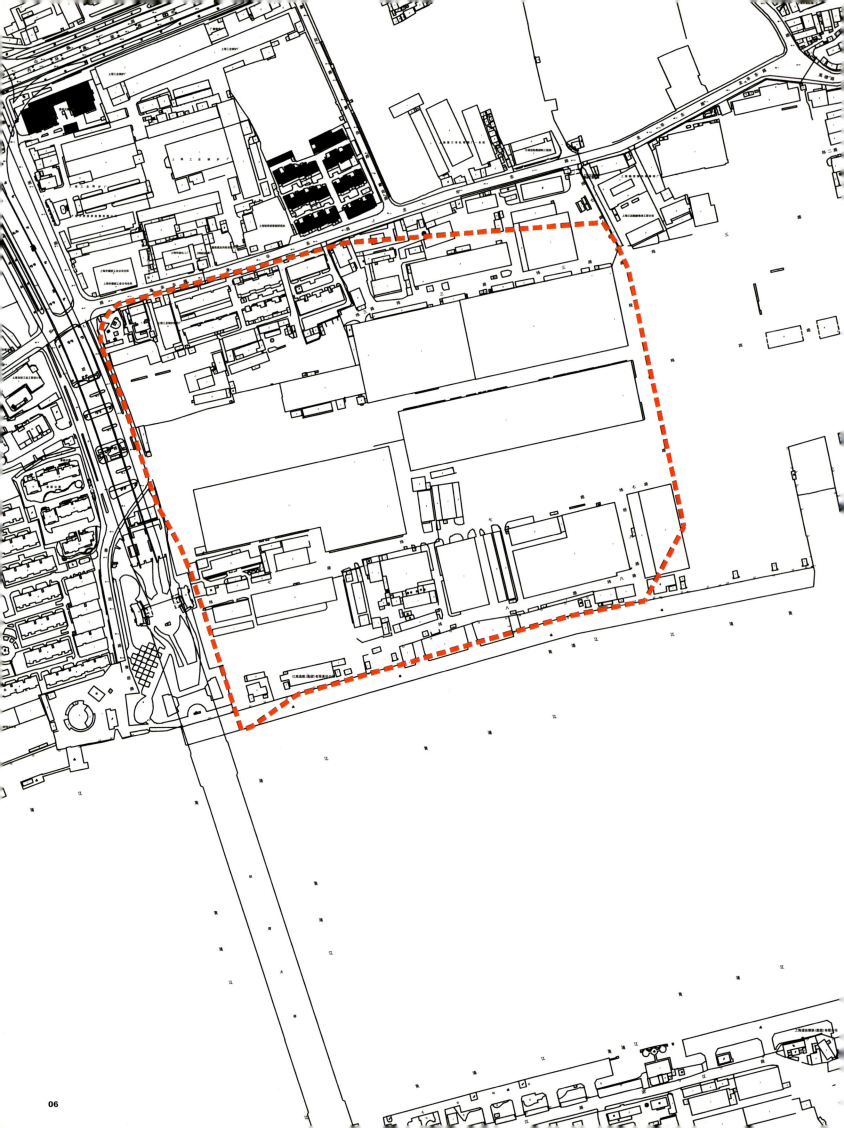

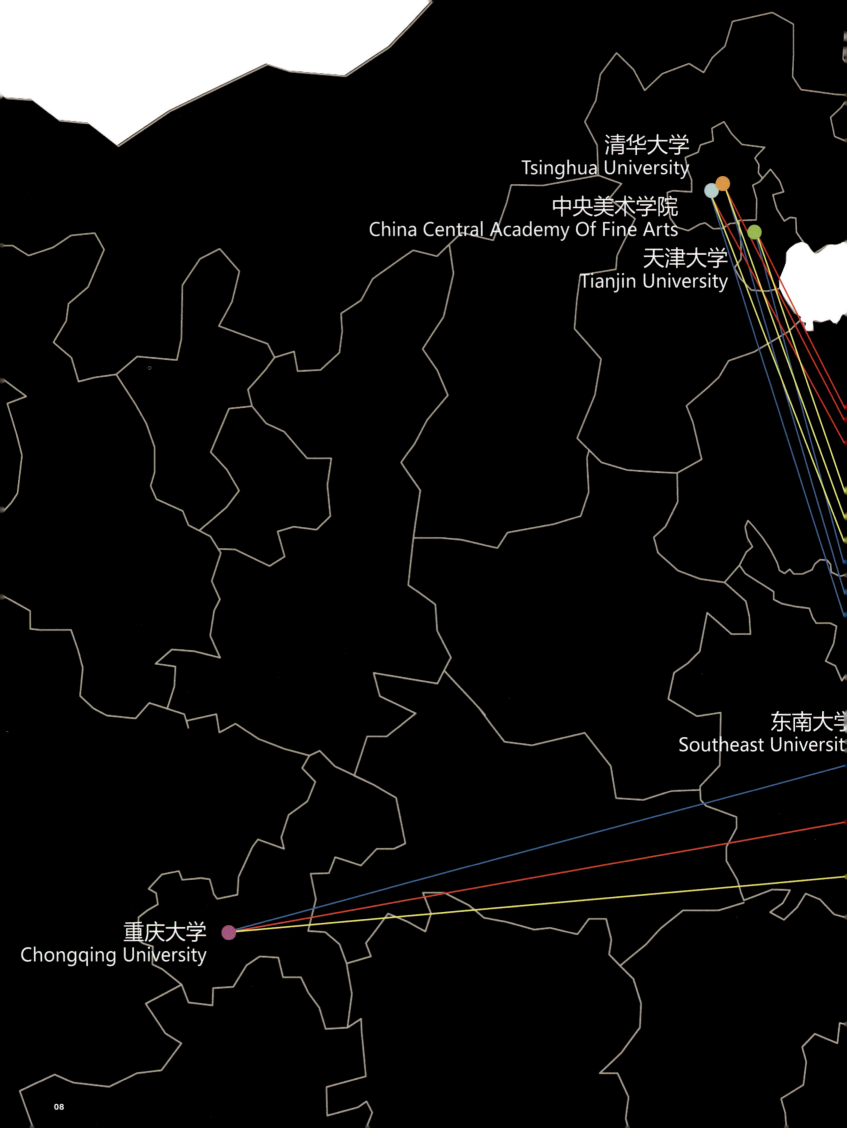

## "七校联合设计" 大事记

同济大学作为这次"七校联合设计"的主办学校，一共组织了三次集中的活动。

第一次是在上海，为期两天。主要组织各校师生在江南造船厂的基地上进行调研活动，并且参观了上海在历史建筑改造方面比较有名的实例。

第二次集中活动在东南大学进行，各校师生带着各自的工作成果在东南大学进行了第一次集中评图。

最终评图安排在浙江大学进行，最终的评图活动非常精彩，各校都在此次活动中得到了收获，在最后的评图发言中，联合设计的教学模式得到了高度的评价。

Tongji Unversity preside over the United-Architectural Design this term,he organize three times of gathering .

Shanghai was the first time. Seven schools,for about 100 students and 20 teachers, from all over china got togather to have a investagation on the site, the west zone of Jiangnan Shipyard company ,what is more,they were leaded to several famous case on historic architectral preservation.

The Mid-Gathering was in Southeast University,Nanjing. It was the forst time when all of the designs were putting togather to have a whole presentation and comment.

The Finall presentation was in Zhejiang University,Hangzhou. It was really a wonderfull gathering, Teachers and Students make progress by this kind of design course which was highly appraised.

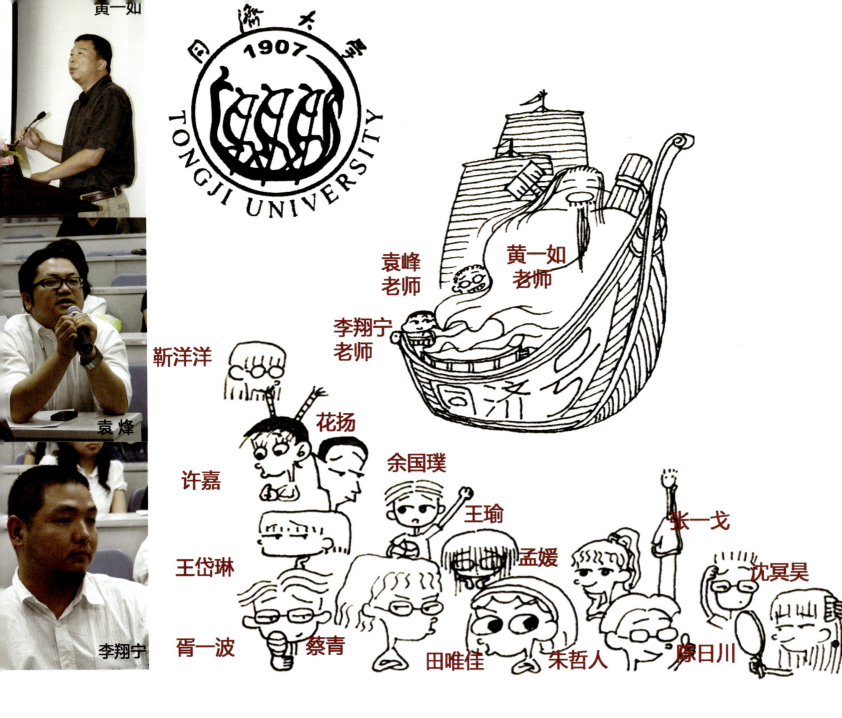

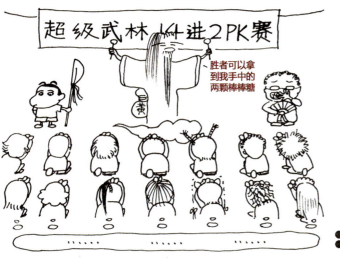

几番混乱后，14人终于统一成三个大组，即尺度蒙太奇，i-industry与印章组。

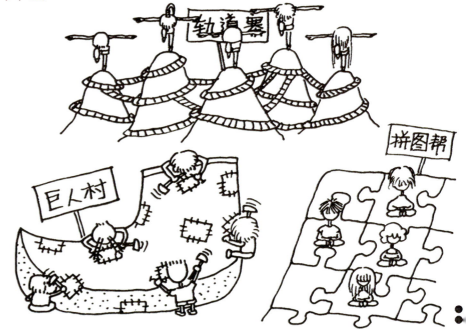

一个月后，七校的中期汇报在莫愁湖畔召开。总的来说，大会很HIGH很成功……

然好景不长，五月初五，专教飘过一朵阴云，只听得云端传出三声尖笑，黄老师显形道——

徒儿们，为了城市设计实施现实过程中的博弈。三个大组各自设计一个城市设计导则。随后把地块重新划分。而每个人则打乱认领他组的地块，依据上家的导则进行单体设计。

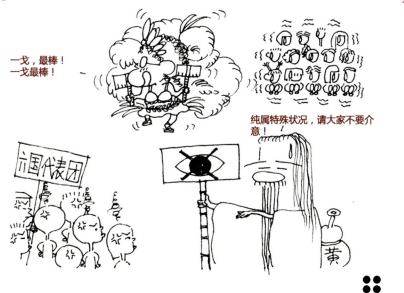

导则并不是一蹴而就的。三大组只好闭门造车，冥思苦想什么导则才能制下家实现我们概念又给予他们足够的自由。

经过两周上下家激烈的谈判，三组的城市设计导则终于完成。接下去的任务自然是分地，场面自然也是极其混乱……

基地被大卸十四块，同学们回自己要设计的地块，开始顶着一个月期限的，踏上了漫漫设计之路

尺度蒙太奇

胥一波　　　花扬　　　许嘉　　　靳阳洋　　　王岱琳

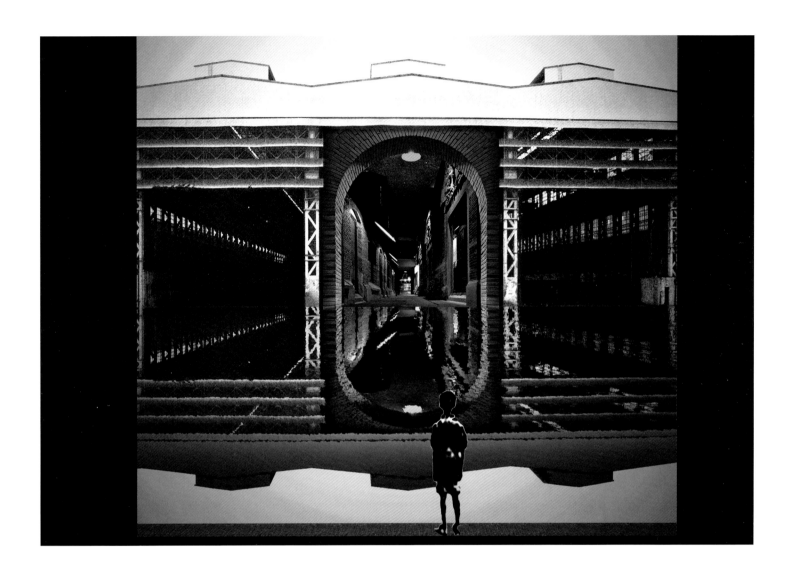

 同济大学 ⟩尺度蒙太奇⟩ 设计：胥一波 花扬 许嘉 靳阳洋 王岱琳
Tongji University >Montage on Scale

# 尺度蒙太奇 Montage on Scale

2010世博会址江南造船厂改造  Expo2010 Jiangnan Shipyard Reservation

1、尺度概念　　　面对这块存有工业历史的第二自然环境，为体现自然界本身的震撼与工业带来的人造自然的巨大差距，我们引入了尺度的概念。在中文中，尺度既包含一种相对的量也包含一种绝对的量，在我们的设计中指的是尺度的变化对比。

2、蒙太奇概念　　　同时为了创造不断地反观自我、不断思考人与自然界的关系的机会，我们又引入蒙太奇的概念。它作为组织空间的一种连接方式，通过快速切换场景，使得强烈的空间印象深刻留存在人们的记忆中。最终的概念结合，我们称之为尺度蒙太奇。

3、城市设计导则　　　由三方面构成，即空间尺度变化、展品尺度变化和蒙太奇。空间变化导则关键是界面的转换，寻求空间尺度的对比，而非一个单一空间的大或者小。展品变化导则是由对比产生的，可以是和人，即参观者的尺度对比，可以是展品之间尺度的对比，也可以是和展品所处展览空间的尺度对比。蒙太奇导则希望这块地上的设计能像电影的蒙太奇手法一样，将一系列分切的尺度变化的镜头组接起来，不断地表现镜头间强烈对比的时空关系，从而加强设计的戏剧性效果，给参观者以震撼的感受。

1.Concept of Scale: Faced with the long industrial history of this site, we introduce the concept of scale in order to present the shocking difference between the industrial environment and the Nature. In our design, scale is an absolute quantity as well as a relative quantity. We adopt the contrast between changing scales.

2.Concept of Montage: Meanwhile, in order to provide an opportunity to re-think ourselves as well as the relationship with human and Nature, we introduce the concept of montage. It is a way to connect and organize spaces, namely to change the scenes swiftly and give the vvisitor a strong impression of space image.
With these two concepts above combined, we have the phrase "montage on scale".

3.Guideline on Urban Design: Guideline is composed of three parts; namely changing scale of space, changing scale of exhibits and montage. The focus of space guideline is changing interface and contrast on space scale. Contrast does not come from the absolute size of a space.Guideline of exhibits is based on contrast with visitors, contrast with other exhibits or contrast with exhibition space where exhibits is placed.Montage guideline means to combine various separate scenes with the means of montage in a movie. This combination would bring strong contrast on time and space, thus give the visitor a drastic feeling.

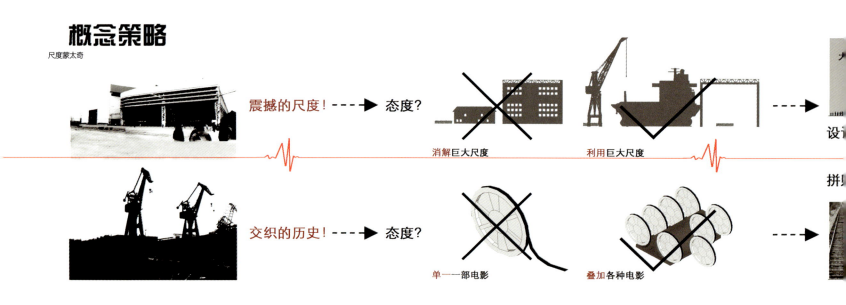

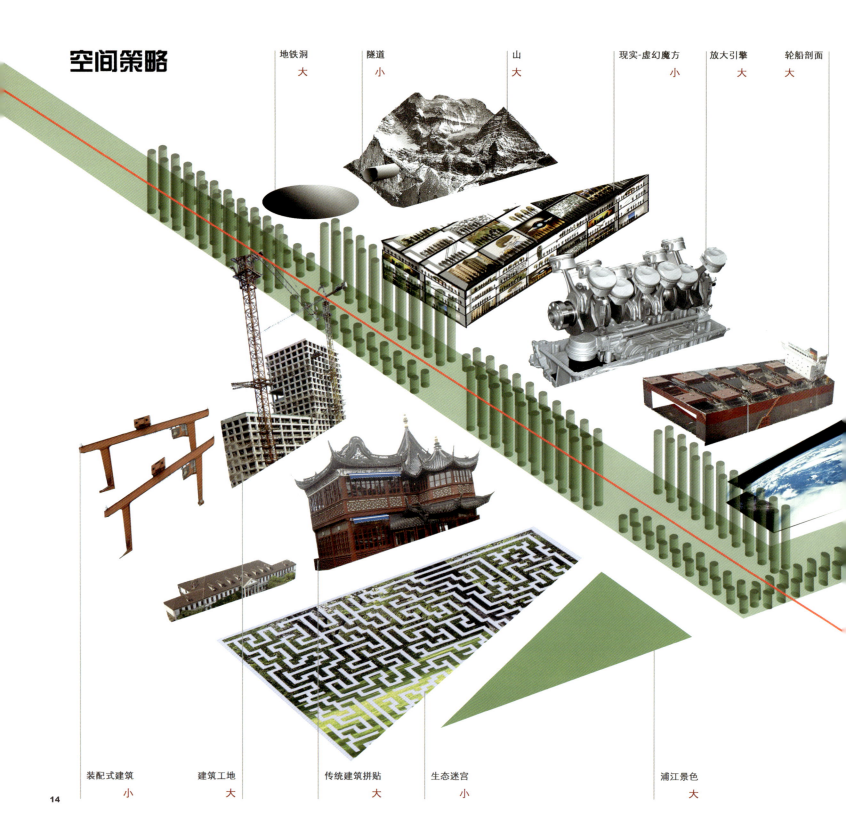

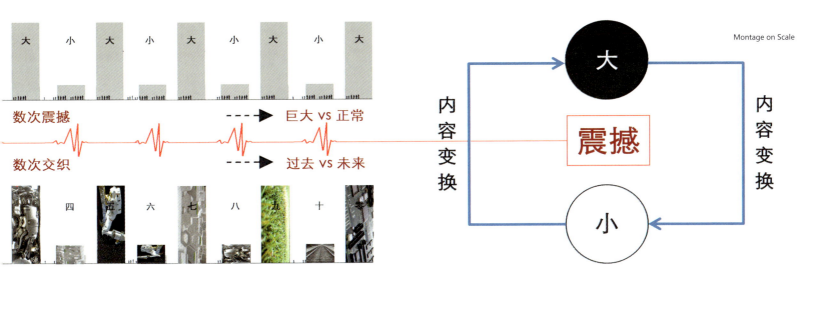

应用于江南造船厂改建的城市设计总原则

目录
1　阐述名词暨设计方法
1.1 尺度
1.2 蒙太奇

2　设计原则
2.1 空间尺度变化
2.2 展品尺度变化
2.3 蒙太奇

3　系统设计原则
3.1 公共空间中，尺度，主题的变化
3.2 绿化景观中，尺度，主题的变化
3.3 建筑外观，整体立面中尺度的变化
3.4 规划需要整洁统一的平面

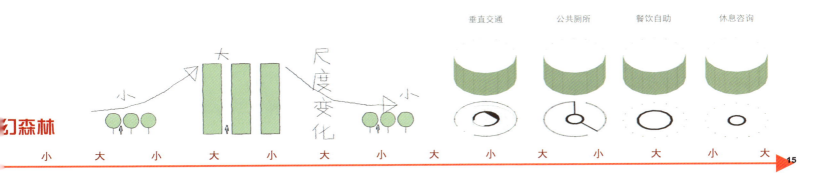

尺度蒙太奇

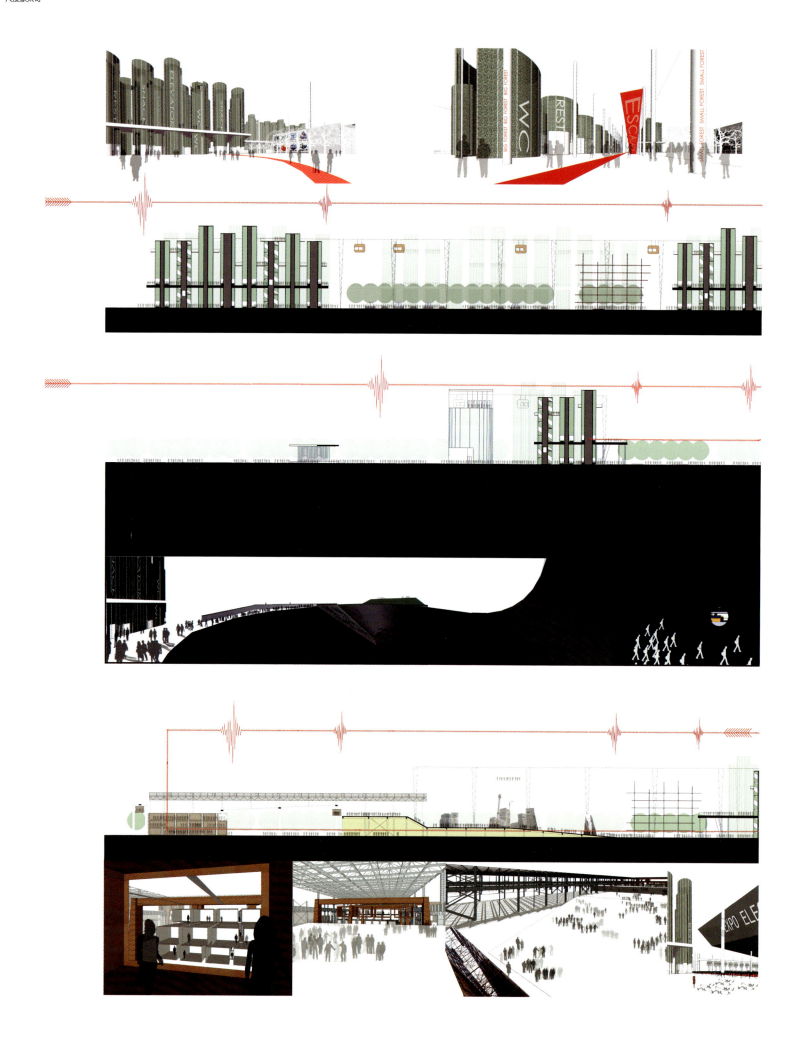

Montage on Scale

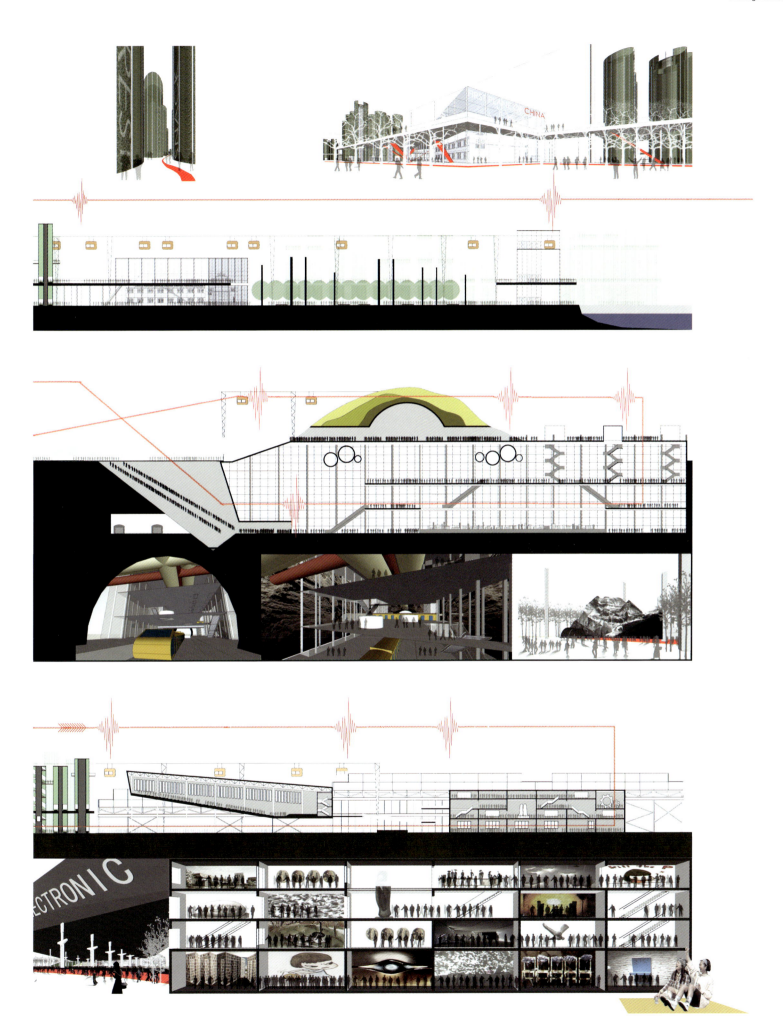

17

尺度蒙太奇

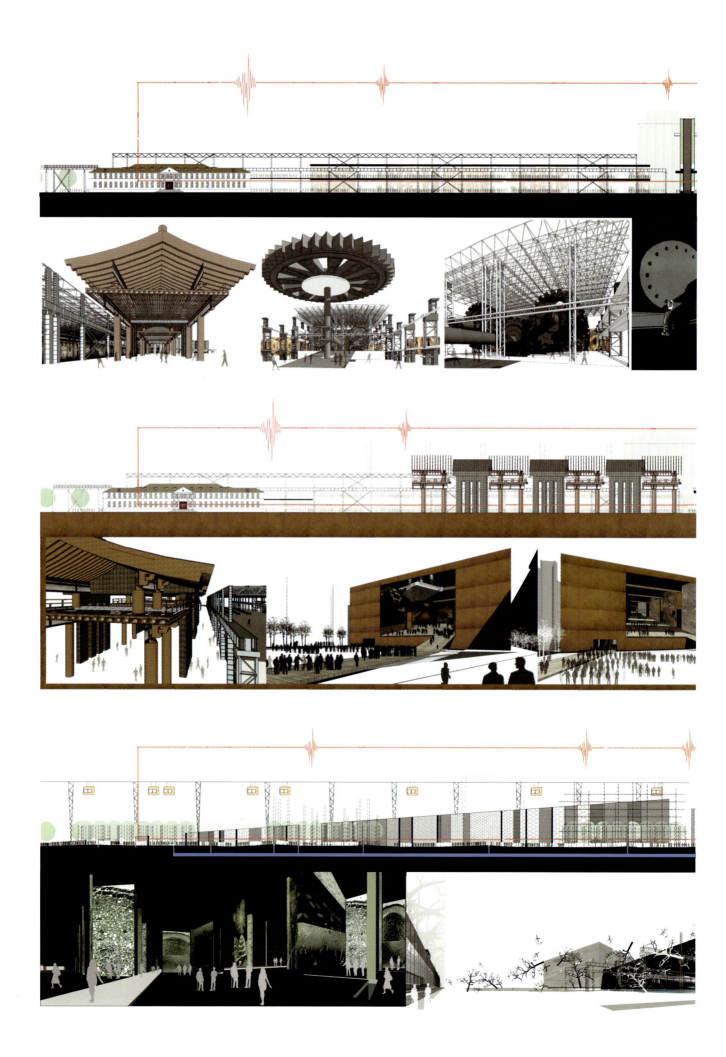

Montage on Scale

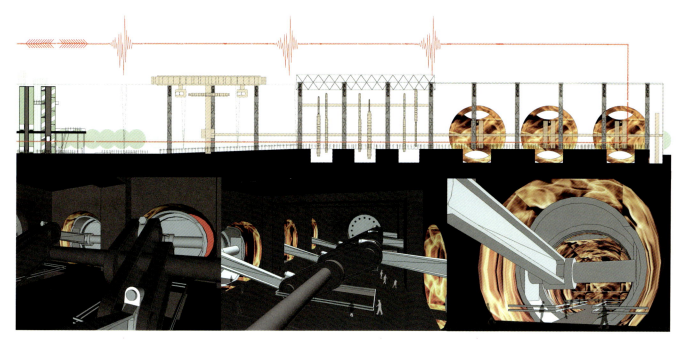
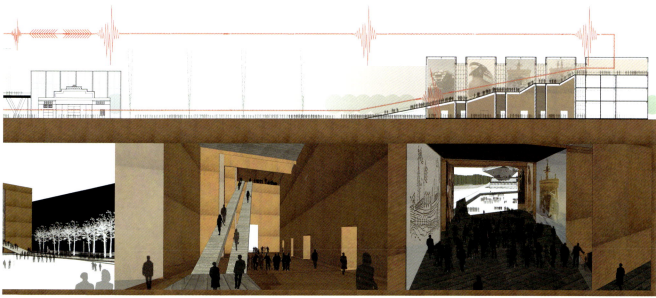
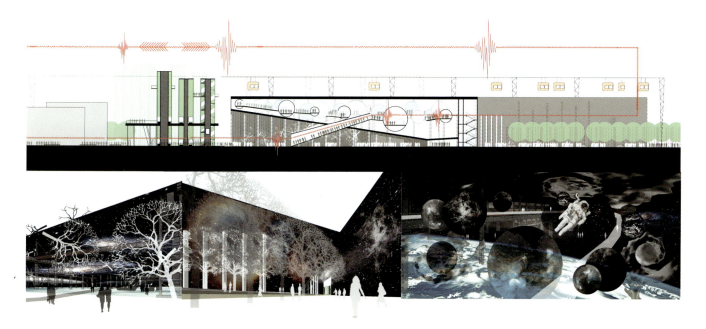

尺度蒙太奇

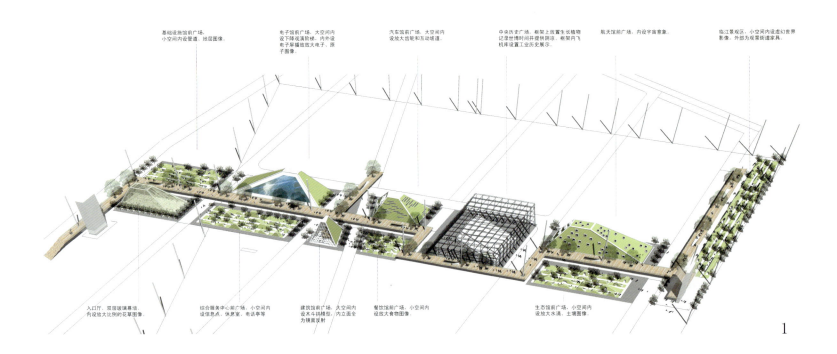

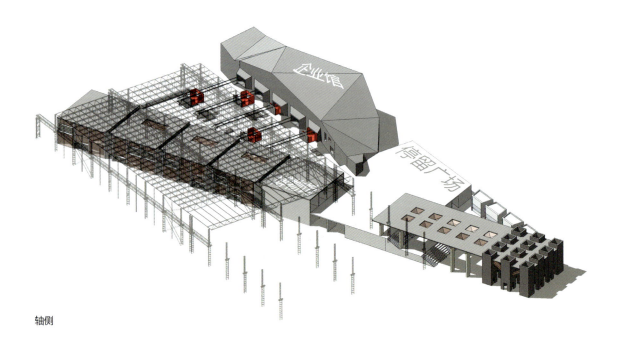

轴侧

1. 主轴设计
2. 建筑产业馆轴测
3. 航空航天馆剖面
4. 基础建设馆剖面
5. 汽车产业馆剖面
6. 建筑产业馆剖面
7. 主轴剖面

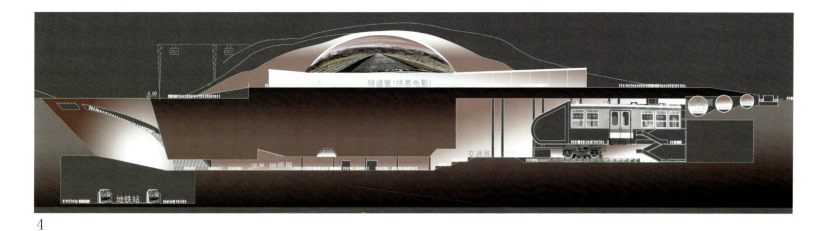
4

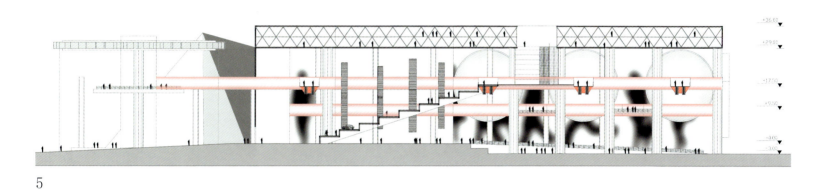
5

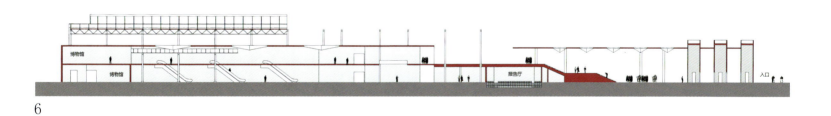
6

7
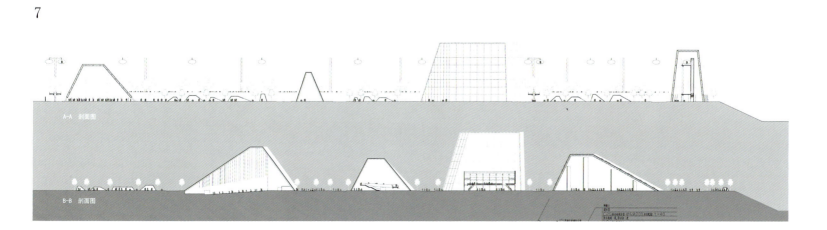

城市印章

印章，代表着权属和记忆，神圣而不可侵犯。
Seal, on behalf of the ownership and memory, is sacred and inviolable.

城市，同样拥有自己的印章。
City also have their own seals.

城市印章,表明地块历史权属,记录城市发展历程，须尊重与保护。
City Seal , which record block authority and urban history, should be respected and preserved.

印章，可分为 阴刻 和 阳刻 ，
Seal have tow kinds of carving, negative one and positive one,

城市印章亦然。
So does Sity Seal.

2008 AD

2010 AD

2050 AD

2100 AD

2500 AD

3000 AD

| 孟 媛 | 朱哲人 | 张一戈 | 陈日川 | 沈旻昊 |
|---|---|---|---|---|
|  |  |  |  |  |

 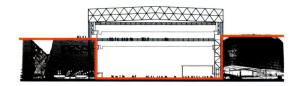

威尼斯城市印章：圣马可广场（阴刻）
CITY SEAL A: PLAZA S.MARCO(NEGATIVE)

巴黎城市印章：凯旋门广场（阳刻）
CITY SEAL B: PLAZA Lion Arc (POSITIVE)

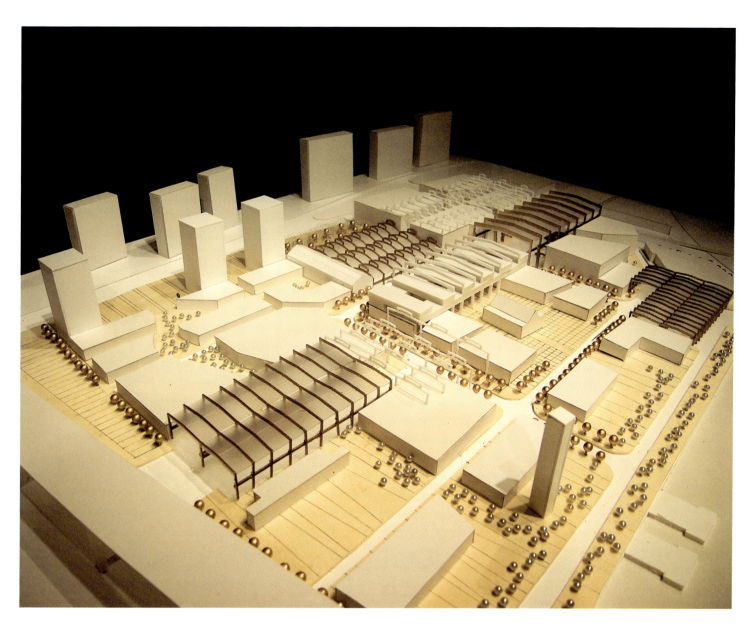

 同济大学 〉城市印章 〉设计:张一戈 朱哲人 沈文昊 陈日川 孟媛
Tongji University > City Seal

# 城市印章 City Seal

2010世博会江南造船厂城市设计  Urban design of the Jiangnan Dockyard for EXPO 2010

1. 保持大尺度空间特征,厂房由实体转换为虚空。
2. 保存工业历史记忆,暴露厂房结构。
3. 保护修缮历史建筑,周边开辟纪念性广场。
4. 保证完整连续的城市公共空间界面。
5. 交通规划适应城市结构,世博会后基本延续不变。
6. 地块划分完整,利于世博会后二次开发。
7. 厂房空间满足世博基本功能,兼顾会后可能功能,打造场所持续活力。
8. 厂房的处理方式尽量减小对地块使用的影响。

1. Keep the factory buildings void as plazas to maintain the sensible super-scale
2. Expose the structure to recall the industrial momery
3. Rehabilitate the historical building and creat square dominated by it
4. Keep the defination of the public space clear and consecutive
5. Planing of roads adapt to the situation during and after expo to make little chages
6. Ensure the area diveided by main street integrity and easy to exploit
7. Schedule the fuctions during and afer expo to ensure the constant vitality
8. Deal with the factory buildings carefully to dectease the influence to environment

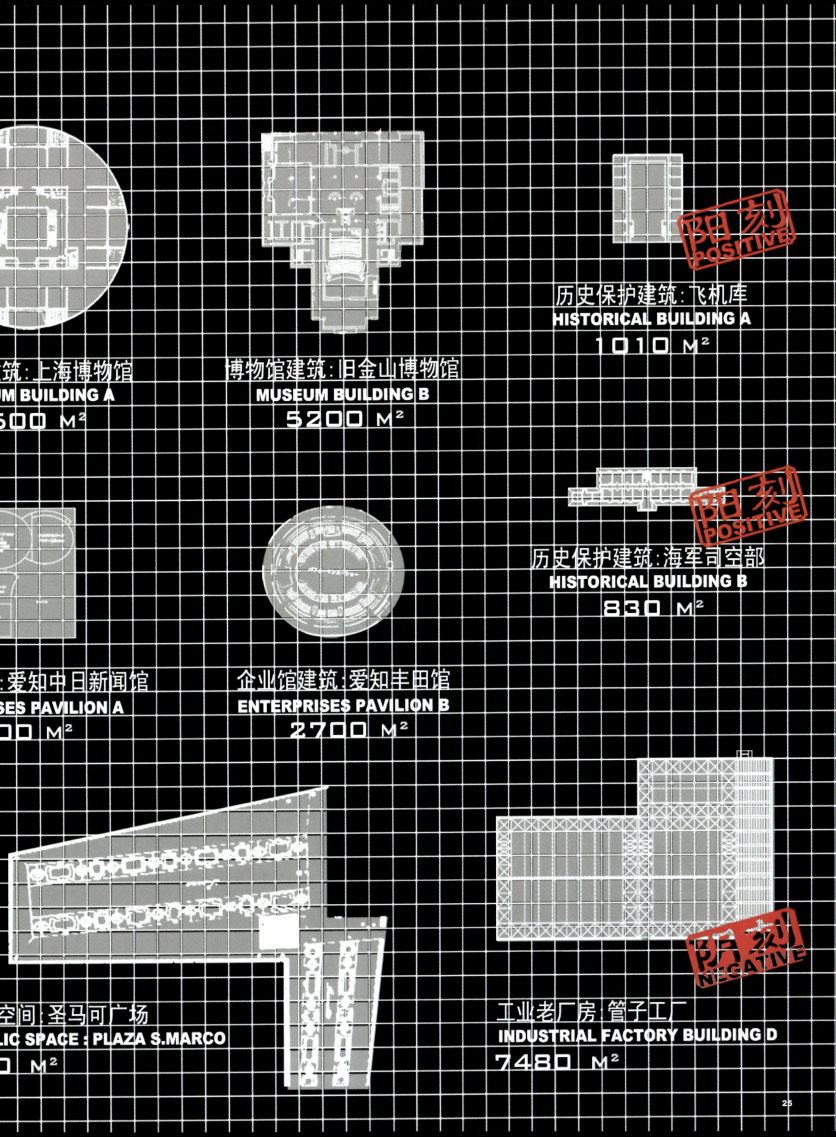

城市印章

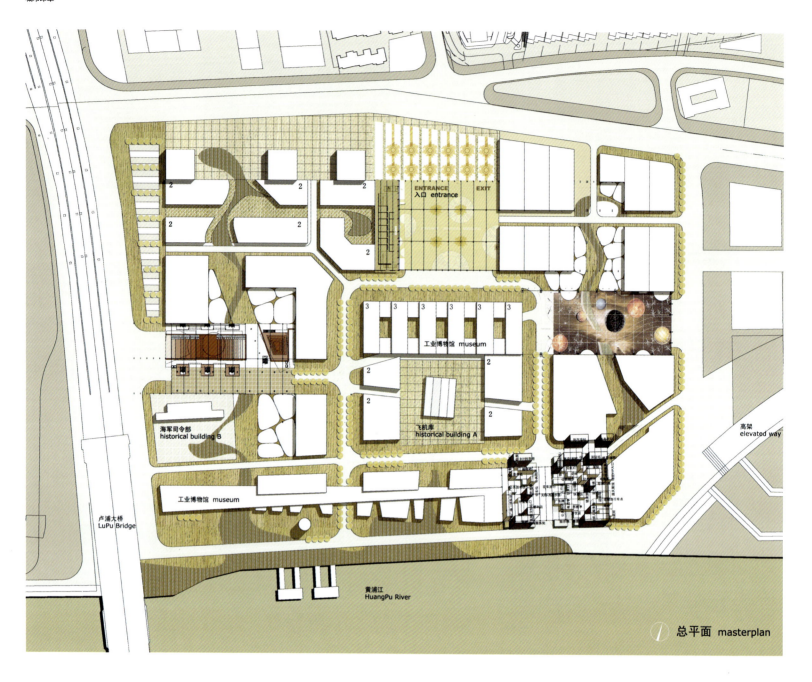

总平面 masterplan

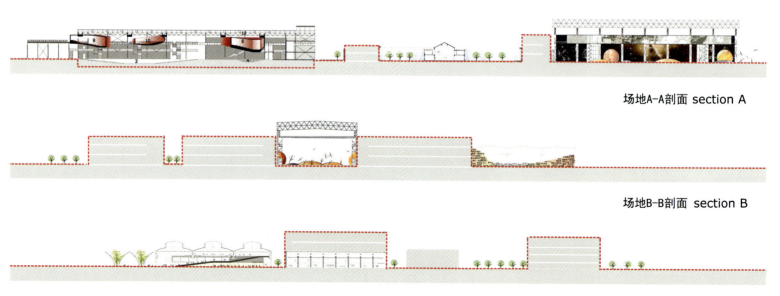

场地A-A剖面 section A

场地B-B剖面 section B

场地C-C剖面 section C

**地块划分 Blocking**
地块划分原则：保证道路分隔的地块的完整性，便于世博会期间展览馆的布置，同时保证世博会之后地块二次开发的便利性与可行性。地块的划分同时兼顾厂房位置，避开厂房使之在地块变革时仍能延续生命。

**功能分区 Function**
功能布置同时兼顾世博会展览需求和厂房保留原则，企业馆分块符合企业入驻数量要求，中心地块作为工业博物馆用地，在世博会之后加以保留。两栋保护建筑周围以设置广场的手段来实现其纪念性。

**形态及功能 Form & Funtion**
在建筑之中建议设置中庭和通过空间，虚实结合完成建筑形态的塑造。图示中为宣布入驻世博会的中国九个企业/行业，大致按类别将其分类，围绕厂房区布置。

**保留与改造 Preservation & Renovation**
厂房保留区，保留结构与围护，保存工业记忆和场所感，形成"阴刻"空间。结构保留区，选择性的保留柱子和桁架，通过再生性利用的方式使其生命在其他建筑中得以延续和保留。

**建筑高度 Heights**
高度布置主要原则：南低北高，中心高度统一创造围合广场，厂房旁建筑低于厂房牛腿高度。滨江标志物高度显示其指向性和标示性。

**开发时序 Delevopment**
世博会之后该园区仍然是城市的组成部分。除了中心区和部分江边的工业博物馆在会后得以保留之外，其余地块的企业馆将被拆除置换为城市新型功能。

**道路结构 Road Net**
路网的设置考虑的是如何在会后直接融入城市肌理之中，所以道路级别与划分方式都要适应城市尺度，对应城市功能。地块主路在会中作为主要人行环路，直接联系四个厂房空间，支路则是考虑捷径系统和货运后勤系统。

**消防 Fire Control**
采用路网和广场相结合的方式，所有的厂房均可做外消防广场来适用。消防体系做到将地块划分成适合扑救的尺度。

**服务与入口 Service & Entrance**
整体地块的服务点需要服务半径的无遗漏覆盖，后勤服务入口和各场馆根据货运系统排布。入口则兼顾主要人行道路位置，同时，这三点都要考虑与厂房的位置关系，依附而互不干扰。

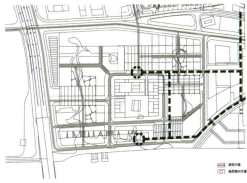

**高架步道 Elevated Pass**
高架步道是世博会园区整体规划中重要的组成部分，而江南造船厂基地区域整体园区的西北角，是入口园区也是高架步道西部的末端，在中心区域设置两高架垂直连接点，并且设计次级高架道路，形成捷径。

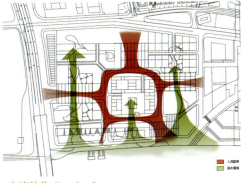

**人流趋势 People Stream**
在地块中心区设置环道，所有厂房围绕安排，辐射性的路网将整体组织起来，而三条景观步道从江滨渗透到基地内部。

**货运后勤 Service System**
作为后勤系统，便捷性成为主要的控制条件，利用道路系统，同时不干扰主道路，形成一个完整的便捷的货运路网系统。

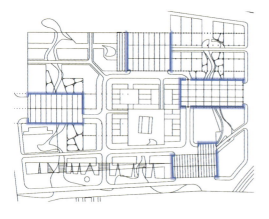
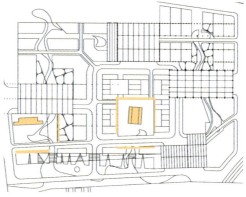
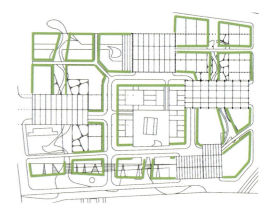

## 工业老厂房改造导则 DESIGN GUIDELINES AND DIAGRAM FOR INDUSTRIAL WAREHOUSE

| | 结构(STRUCTURE) | 面积(AREA) | 高度(HEIGHT) | 长度(LENGTH) | 宽度(WIDTH) | 退界(SETBACK) | 开口(HOLE) | 视线(VISION) |
|---|---|---|---|---|---|---|---|---|
| 屋顶(ROOF) | A1 | A2 | A3 | A4 | | A6 | A7 | |
| 地面(GROUND) | | B2 | B3 | B4 | B5 | B6 | | |
| 立面一（面对主要出入口）(ELE 1) | C1 | C2 | | | | C6 | C7 | |
| 立面二（邻企业馆）(ELE 2) | D1 | D2 | D3 | D4 | D5 | D6 | D7 | |
| 内部空间(INSIDE SPACE) | | E2 | E3 | | | | | E8 |

## DESIGN GUIDELINES AND DIAGRAM FOR PROTECTED BUILDINGS 历史保护建筑改造导则

| | 结构(structure) | 表面(surface) | 高度(height) | 宽度(width) | 空间(space) | 退让(recede) | 开口(hatch) | 绿化(vegetation) | 其他(others) |
|---|---|---|---|---|---|---|---|---|---|
| 保护建筑主立面(facade) | A1 | A2 | A3 | A4 | | A6 | A7 | A8 | A9 |
| 保护建筑侧立面(side elevation) | B1 | B2 | B3 | B4 | | B6 | B7 | B8 | B9 |
| 屋顶界面(roof) | C1 | C2 | C3 | C4 | C5 | | C7 | C8 | |
| 内部空间(inside space) | D1 | D2 | D3 | | D5 | | D7 | D8 | |
| 广场地面(ground) | E1 | E2 | E3 | E4 | | E6 | E7 | E8 | E9 |
| 广场地下空间(underground space) | F1 | F2 | F3 | F4 | | | F7 | F8 | F9 |
| 广场周边界面(interface) | | G2 | G3 | | | G6 | G7 | G8 | G9 |

## 历史保护建筑改造导则 DESIGN GUIDELINES AND DIAGRAM

| 世博用(2010 expro use) | 世博后用(after 2010) | | 路宽(width) | 人行通道宽(width) | 建筑退界(recade) | 路面标高(height) | 通过方式(way) | 视线遮挡(view) | 街道家具(furniture) | 绿化植物配比(plant) |
|---|---|---|---|---|---|---|---|---|---|---|
| | | 世博园内主要交通空间 (main road) | A1 | A2 | A3 | A4 | A5 | A6 | A7 | A8 |
| | | 城市主要交通空间 (main road) | B1 | B2 | B3 | B4 | B5 | B6 | B7 | B8 |
| | | 世博园内次要交通空间 (subaltern road) | C1 | C2 | C3 | C4 | C5 | C6 | C7 | C8 |
| | | 城市次要交通空间 (subaltern road) | D1 | D2 | D3 | D4 | D5 | D6 | D7 | D8 |
| | | 世博园内景观通廊 (view corridor) | E1 | E2 | E3 | E4 | E5 | E6 | E7 | E8 |
| | | 城市景观通廊 (view corridor) | F1 | F2 | F3 | F4 | F5 | F6 | F7 | F8 |
| | | 广场及其他公共空间 (other public space) | G1 | G2 | G3 | G4 | G5 | G6 | G7 | G8 |
| | | 广场及其他公共空间 (other public space) | H1 | H2 | H3 | H4 | H5 | H6 | H7 | H8 |

公共空间限定：
阳刻 / 阴刻
新 新建筑(new) / 空 空地(plan)
保 保护建筑(protected) / 厂 厂房(warehouse)
架 架高(elevated) / 下 下沉(descend)

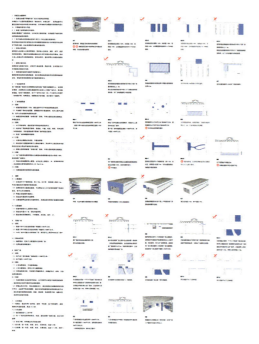

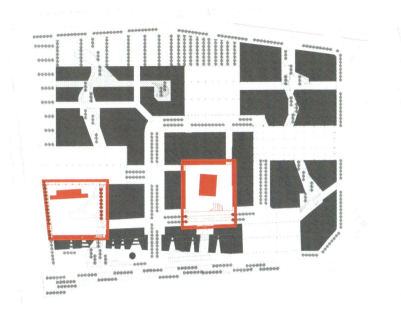

历史保护建筑——正常尺度——统治广场——城市印章——阳刻
HISTORICAL BUILDING — NORMAL SCALE — PALZA DOMINATED — POSITIVE

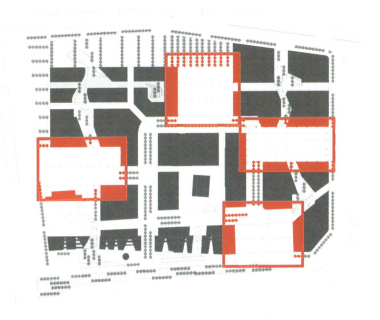

老工业厂房——超常尺度——围合广场——城市印章——阴刻
FACTORY BUILDING — SUPER SCALE — PALZA ENCLOSED — NEGATIVE

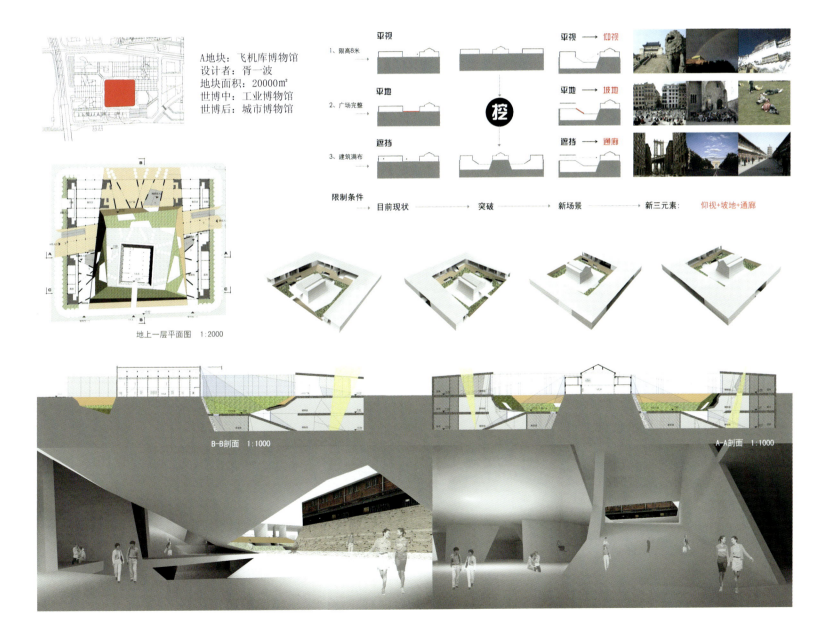

城市印章

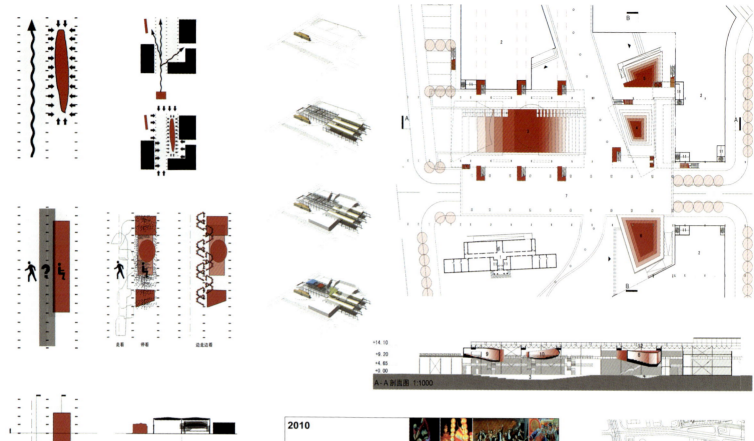

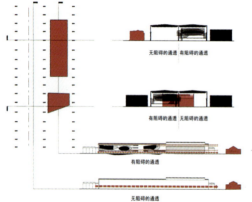

B地块：部件焊装车间
设计者：田维佳
地块面积：7800m²
世博中：庆典广场
世博后：城市剧场

C地块：西区装焊车间
设计者：花扬
地块面积：8600m²
世博中：登月计划
世博后：体育健身

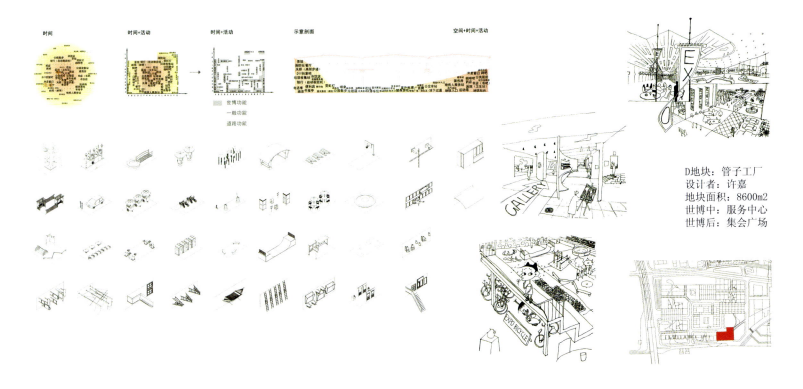

D地块：管子工厂
设计者：许嘉
地块面积：8600m2
世博中：服务中心
世博后：集会广场

E地块：船体联合车间
设计者：蔡青
地块面积：8200㎡
世博中：入口广场
世博后：城市绿地

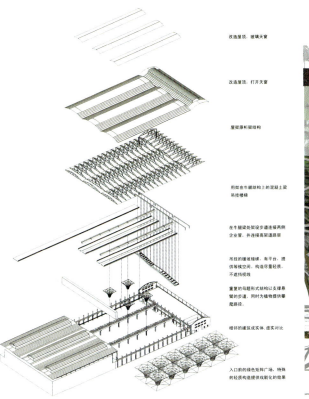

# i-industry 城市设计

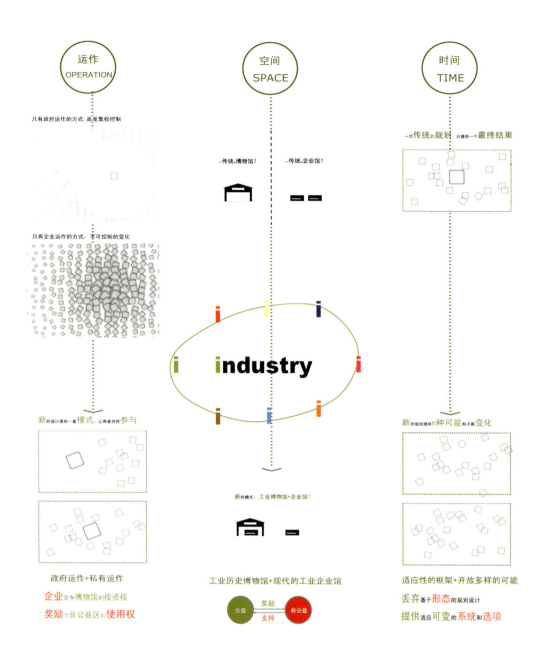

余国璞

蔡青

田唯佳

王瑜

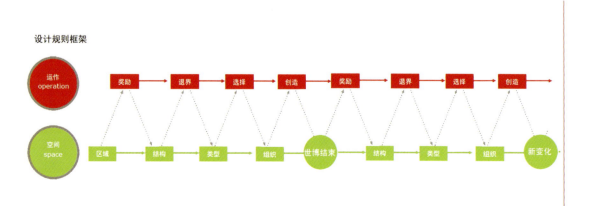

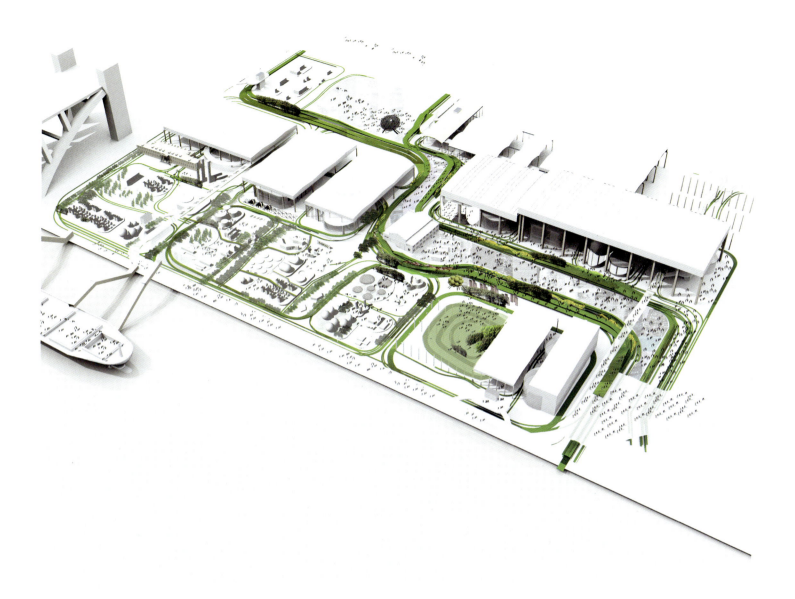

同济大学〉i-industry 城市设计〉设计：蔡青 余国璞 田唯佳 王瑜
Tongji University〉i-industry Urban Design〉Group Members : Cai Qing / Yu Guo Pu / Tian Wei Jia / Wang Yu

# i-industry 城市设计 i-industry Urban Design

2010世博会址江南造船厂改造  EXPO2010 Jiangnan Shipyard Reuse

　　江南造船厂——上海2010年世博会工业博物馆园区改造是一个复杂而多元的项目。在时间上，世博会赋予这块基地的意义，是终结，还是一个新的开始？在空间上，建造博物馆还是企业馆，将两者并存还是分离？在运作方式上，是采用政府主持单一运作，还是完全开放的私有化运作？众多矛盾在时间，空间，运作三个方面涌现，所以我们决定，新的城市设计策略就从这三个方面展开。

　　我们提出i-industry这一概念，主张一种动态的城市设计方法，有众多的参与者，遵循着相应的规则，产生着弹性的变化。在时间层面上，摒弃追求得到具体的设计结果，取而代之的是提供一个适应性的框架，关注变化的过程；在空间层面上，模糊不同展示性质，将工业历史博物馆和企业馆融合起来；在运作层面上，主张建筑师抛弃惯有的主导姿态，仅扮演协调者的角色，思考怎样吸引更多的参与者，怎样制定规则，怎样运作使得非公益支持公益，公益奖励非公益。"i-industry"中的"i"代表工业的各个门类、个体、私有单位、每一个参与者；"industry"代表整个工业、公共、共有单位、政府和管理者。以奖励、运作来激励、协调个人与公共，让更多的个体能够参与城市更新，即为i-industry城市设计的目标所在。

This project "Jiang Nan dockyard factory rebuilding" ——2010 Shanghai EXPO site and industry museum Renovation is a complicated and multi-focused project. The question goes: is EXPO an end or a new start? Should it be museum only or enterprises pavilion only? Should the government pay the bills only or individuals share. There lie many conflicts which fall into three factors: time, space and operation, thus the new urban design strategy just starts from those.

Dislike a conventional way, i-industry advocates a dynamic way of urban design which encourages multiple actors playing under variety of regulations and flexible variations.As for the TIME factor, i-industry discards the conventional way of a providing a fixed spatial result, but create a suitable frame which brings changeable opportunities.As for the SPACE factor, i-industry blurs the distinction between different exhibition function and integrates industry museum and enterprise's pavilion together as a whole. As for the OPERATION factor, i-industry discards the only desire of designer and redefines the role of designer as a coordinator who organizes the way in which more actors play and ensure that profit supports non-profit and the profit get reward accordingly.In the concept of "i-industry" "i" represents fractions of industry, individuals, profit and every actors; while "industry" represents the whole industry, public, non-profit and government and management. Through the way of rewarding policy, benefits of individuals and public are motivated and balanced, thus allow varieties of actors participate into urban design process. And this is just the goal of i-industry urban design.

i -industry 城市设计

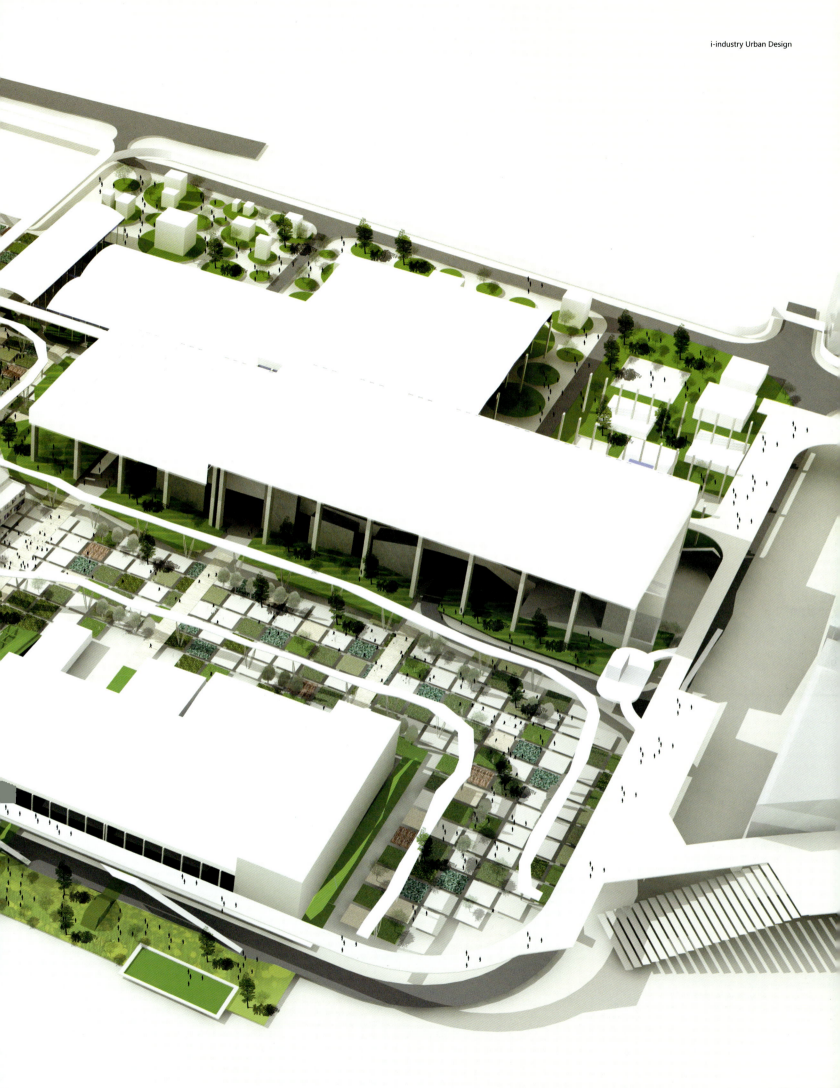

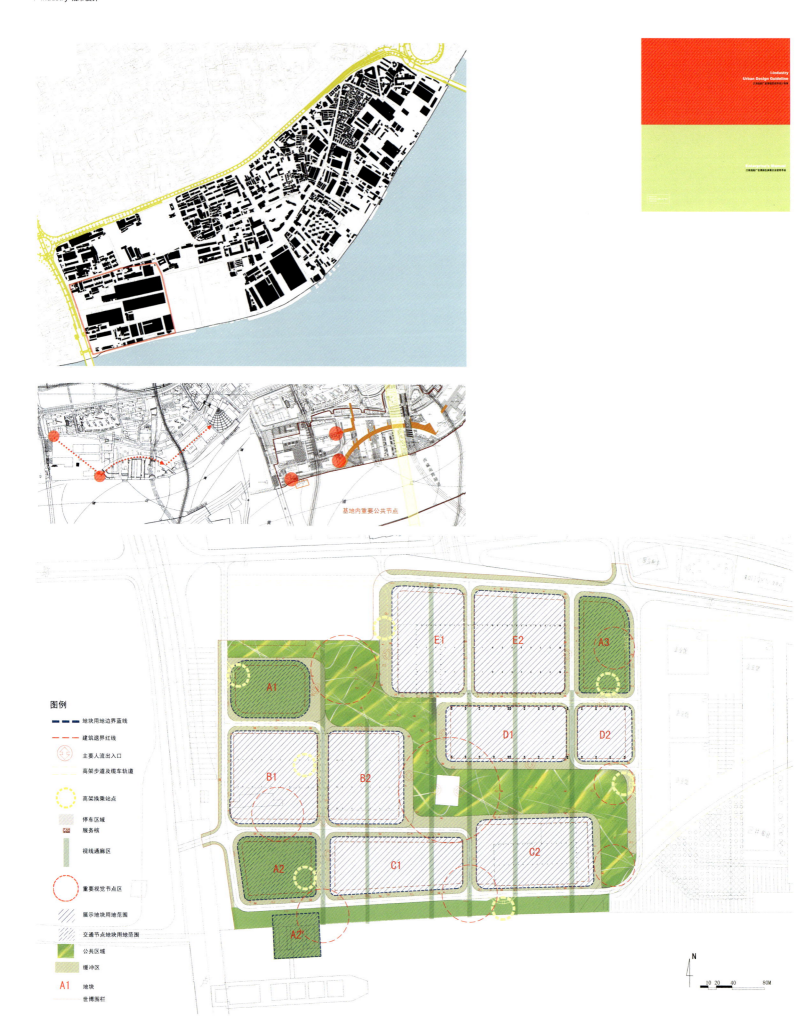

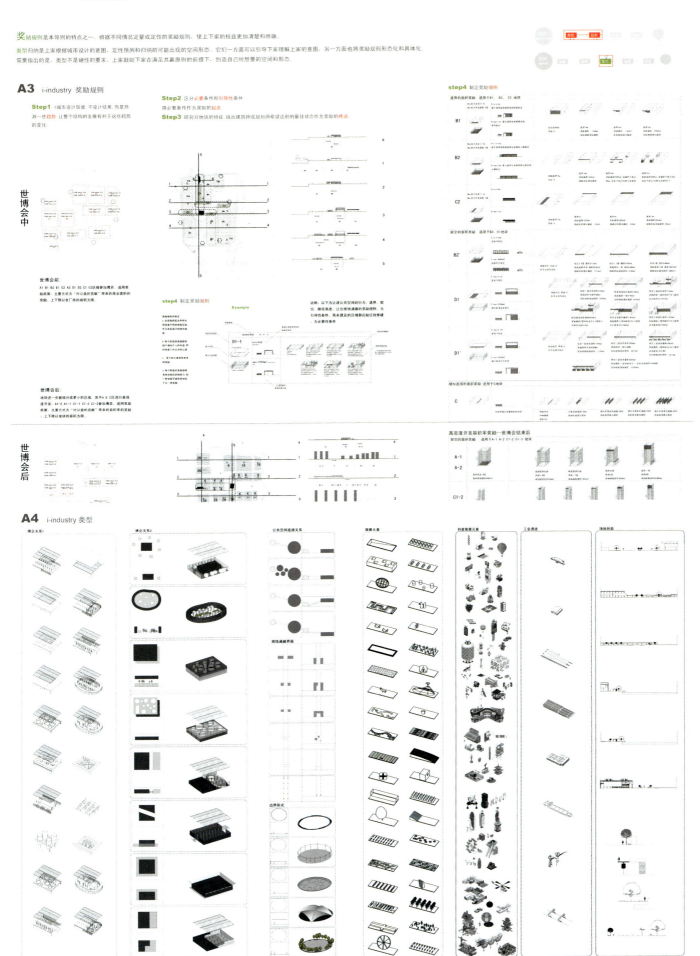

# i-industry 城市设计9原则

**1**

原则1 运作：鼓励更多的参与者

众多的参与者，政府、企业、设计师、参观者、市民等在不同时期共同完成i-industry城市设计。

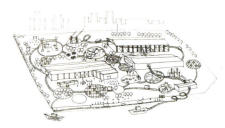

空间：丰富的主题，互动的设计

代表不同利益的个体参与设计
每个部分都有独立的主题和个性

**2**

原则2 运作：由非公益区，公益区，缓冲区组成

非公益区支持公益区建设，同时得到奖励，非公益区与公益区共同建设缓冲区。

公益区指公共的道路和广场，不含有任何私有利益的部分，非公益区是各个企业参与的区域，最大限度的私有性质分配了不同参与者的权限，用缓冲区作为弹性的边界支持和奖励的关系成为参与的动力。

空间：活跃的企业展示区，宜人的公共区和高复合的缓冲服务区。

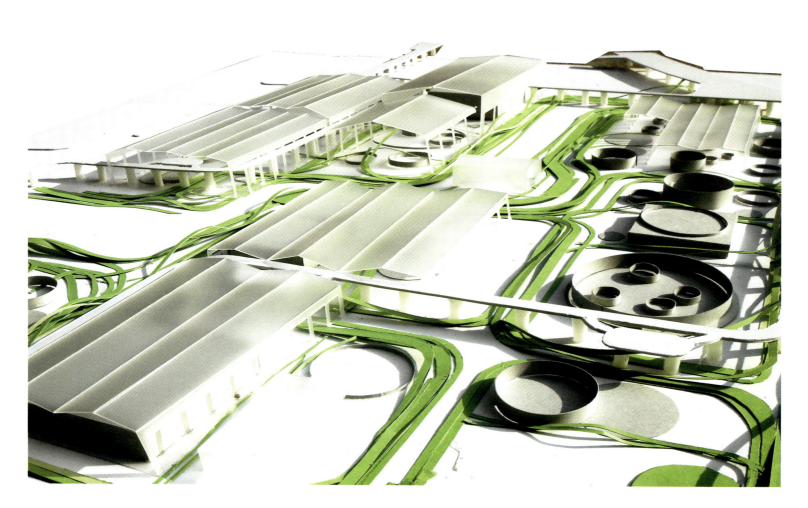

## 3
**原则3 运作：更多的视线连接奖励更大的展示密度**

如果企业的展览场地能够回馈公共更开放的视线，尤其在重要的视线节点之间，在这些视线的通道上，企业将被奖励更大的展示密度。
开放式的展览有利于形成欢乐畅快的气氛，在重要节点之间建立更多的视线交流有利于对空间的感知。应鼓励有利视线开敞的展览方式。
世博会后通廊变成道路，完善城市的功能。

**空间：通透的视线**

## 4
**原则4 运作：越多深入非公益区的道路，奖励更多入口。**

鼓励缓冲区的绿丝伸入非公益区。

**空间：通达的交通**

基地内展示区域获得便利的交通可达性，增加展区影响力。

## 5
**原则5 运作：越开放边界 奖励越多容积和展示面**

封闭的边界带来给定面积的减减，开放的边界带来给定面积的增加

**空间：开放的边界**

基地内公共空间两侧有开放的边界，边界上有丰富的展示内容。

## 6
**原则6 运作：规定地块内越多的退界，奖励越多的容积率。**

鼓励在地块面向公共空间或历史保护建筑的一侧贡献开放空间。

**空间：以历史建筑为中心的广场**

在历史保护建筑附近的视线控制区，鼓励采用积极的退让来营造开放性广场。
鼓励底层采取适当架空或竖向共享空间以增加地块的开放性。

## 7
**原则7 运作：更多的绿地面积，奖励更多的展示密度。**

规定地块内绿地面积不计入计算容积率内，贡献更多的绿色（树木、草地和水体）奖励更多的展示密度和展示形式

**空间：更多的绿色**

将整体的绿化任务分解为不同个人完成形成的结果，这样的方式导致了绿色的印象更频繁地出现在场地的更多的部分。除了集中绿化，点状分散的绿化也无处不在。

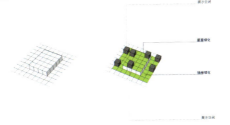

## 8
**原则8 运作：鼓励展示方提供公共服务设施**

非公益区的展示方若在缓冲区提供公共服务设施，可获得公共服务设施的设计权以宣传展示

**空间：充足多彩的公共服务设施**

将整体建设基础设施的任务分解为不同个人完成的结果。不同展示区外边缘的缓冲区内设置了丰富的有特色的公共服务设施，同时服务公益和非公益区，世博会后，大部分的基础设施被保留下来，继续服务城市。

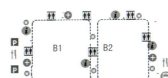

## 9
**原则9 运作：鼓励临时性**

修建临时建筑和使用可回收材料可获得更大的展示面积。临时的建筑带来拆卸的可变性，可回收的材料符合世博会的主题理念。

**空间：变化的可能**

会中修建的临时性展示建筑拆卸后，不影响会后场地的使用，以适应新的功能。

世博前期　Before EXPO 2008.05~2010.03

世博期间　EXPO2010 2010.05~2010.10

世 博 后　　After EXPO 2012.01~2062.01

世 博 后　　After EXPO 2012.01~2062.01

**指导教师**

**许懋彦**
Xu Maoyan

**张 利**
Zhang Li

**卢向东**
Lu Xiangdong

**刘念雄**
Liu Nianxiong

**小组一**

**曹 雩**
Cao Yu

**何 珊**
He Shan

**郑秀姬**
SuHee Jung

**沈 华**
Shen Hua

**乔** 
Qiao Tu

小组二

**梁其伟**
Liang Qiwei

**李 娜**
Li Na

**郑瑄**
Zheng Xuan

**戴 南**
Dai Nan

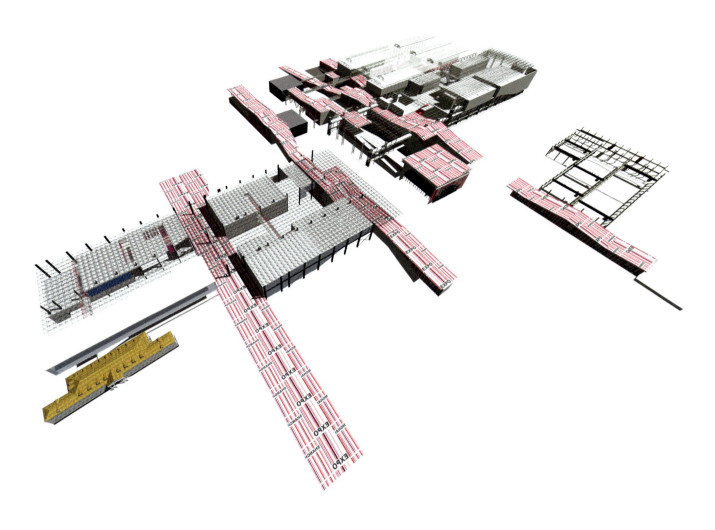

清华大学 ⟩ 城市经纬 ⟩  设计： 晋雯 何珊 郑秀姬 沈华 乔拓
Tsinghua University > Intertexture

# 城市经纬　　Intertexture

舞台：上海世博园区浦西区（原属于江南造船厂工业区）
演员：上海世博会参展企业（上海大众、中国铁路等）为数众多的娱乐、餐饮、商业、游乐等城市商家
导演：清华大学七校联合第一小组
策划：上海世博会
剧情：场景一：厂区已经搬迁，四栋较大的原有建筑厂房与工业构架被保留。主演入场。保留厂房的内部巨大尺度与企业的人性化尺度形成鲜明对比，巨大尺度的空间被按照场地东西纬线方向的结构轴线分割消减为较小的人性化尺度的展览空间。
　　　场景二：由于保留厂房的超大东西长度，南北向城市空间被厂房割裂。地段南北向按场地经线在四个区域被打通，形成南北经线方向的通廊，城市生活空间与黄浦江联系紧密。
　　　场景三：众多附属演员涌入舞台，围绕着南北通廊组成了一组新的城市街道空间"城市综合体"，满足世博会期间需求。以旧结构为基础的东西向展览空间与南北向的城市街道空间穿插形成交叉编织的效果，引导人流南向分流，组织世博观览。穿插过渡空间成为观众注意的重点。
　　　场景四：世博会结束，主演陆续离场。而众多附属演员却成为舞台后续剧段的主演，利用形成的连接城市与黄浦江的街道空间带动两边区域的开发，形成鱼骨式发展。。。
主旨：尺度消减、功能编织
Site: Pu Xi area of the expo park in Shanghai(Used to be an old factory zone)
Cast: Corporations of Shanghai Expo Park(such as Shanghai Masses, China Subway)　Great numbers of urban businesses(entertainment, food, shopping )
Director: Team I of Tsinghua University
Hatcher: : Shanghai Expo Park
Gut: Scene I　Four oversized historical industrial workshops are standing on the stagnant site, after the moving out of the industry. Leading actors enter. For the big contrast be　tween sizes of workshop and people, the huge space is cut into smaller and comfortable one according to the W-E reference lines.
　　Scene II　The dissevered the N-S urban space is got through within four channels through the long W-E historical workshops.
　　Scene III　Other actors come in. Newly built "Urban Complexity" is organized to feed the Expo's needs. Crossing with each other, the Complexity frees the big numbers of people.
　　Scene IV　After the Expo, leading actors are replaced by the other ones, together with the Complexity leading the exploitation of the site..
Lemma: curtailment in size, intertexture in function

城市经纬

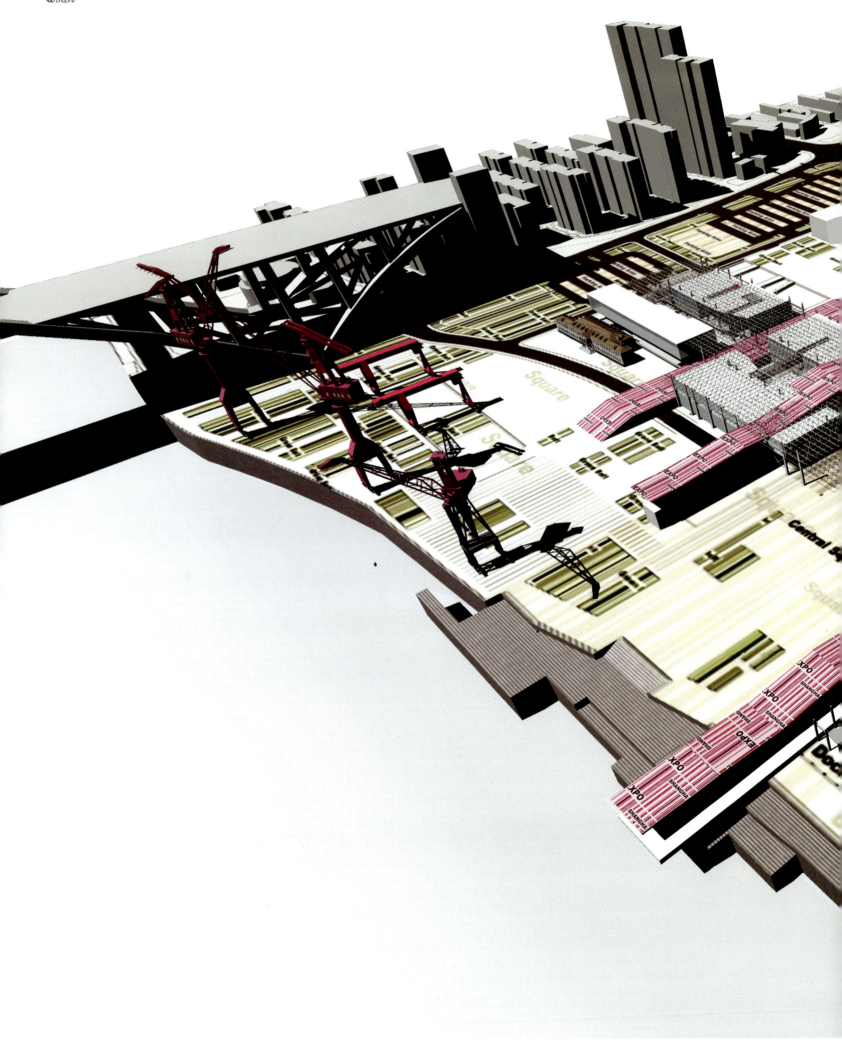

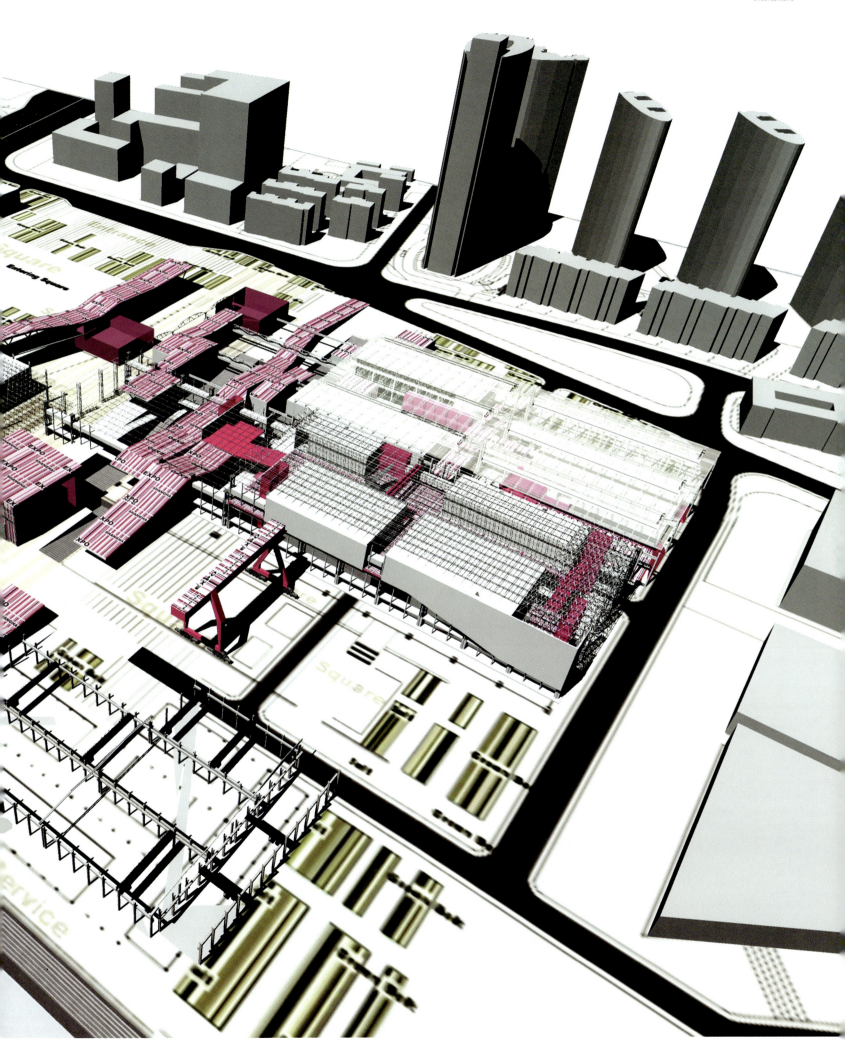

城市经纬

原有厂房尺度分析

世博园浦西区规划尺度分析

地段周边城市肌理尺度分析

城市路网结构尺度分析

我们对场地现状中具有保留价值的厂房的肌理进行尺度分析，并与周边的城市肌理进行尺度对比。工业建筑所具有的大尺度、大体量属性使得每一个大厂房在场地中都占有巨大的空间，因而对场地中的交通限制以及可达性产生了很不利的影响。这种建筑肌理对相对简单的工业生产流程并没有太大的影响，但是在大量人群涌入的状况下，占地巨大的厂房们却成了阻隔入口与沿江地带的巨大障碍。我们提出切割建筑的概念，把原来的大体量切碎成到一个相对小的尺度，打破原来的整体形象。这样，原来的建筑单体变成了几个小建筑体组成的组群，建筑体之间切出来的缝隙产生了新的交通通道，为行人提供了更多样的路径选项。原有的整体大空间通过切割化整为零，产生了多样化的小空间使用可能性。

原有厂房整体大空间

切割·建筑尺度消解·体量分割

复杂丰富多样化的建筑内部空间

切割

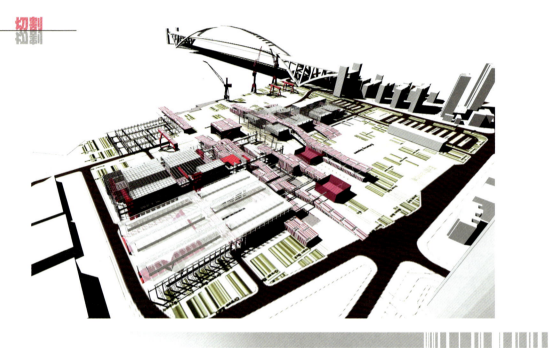

概念 CONCEPT

世博园浦西沿江地带肌理分析

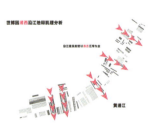
世博园浦西沿江地带肌理分析

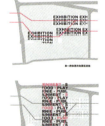

世博园浦西沿江地带肌理分析

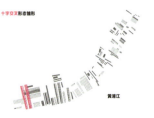
十字交叉形态雏形

分析世博园浦西区沿江地带的建筑肌理，我们发现沿江建筑的走势多以垂直江岸为主导方向，而回顾我们的设计场地，原有的建筑基本是东西向平行于江岸，明显阻隔了北侧沿街的入口区域和黄浦江岸滨水空间的联系。因此我们提出了十字交叉的形态雏形，将相互垂直的两个系统组织到一起，纵横编织形成一个整体。在形态上，纵向肌理的引入很好的延续了浦西区建筑的总体走势，在沿江一线形成一片完整的城市界面；功能上，我们希望在展览职能之外，引入"城市活力综合体"的概念，在世博会的博览主题中加入更多的城市活力因子，通过商业、餐饮和娱乐带动整个区域的城市活力。

编织

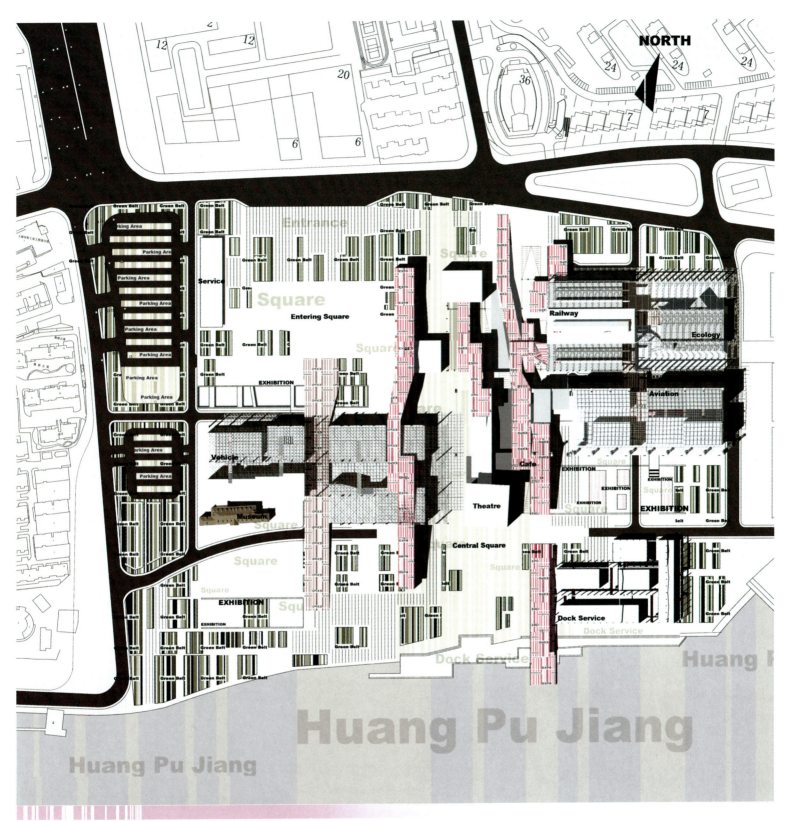

## 总图 SITE PLAN

以建立人性化，生活化场所为总出发点，将尺度的消减与功能的编织做为处理手法，将场地横向上将原有巨型厂房做空间上的消减，同时作为世博会期间展览性功能建筑使用；纵向上打通原来巨型厂房割裂的城市空间，同时编入生活性功能建筑——"城市综合体"（商业，餐饮，休闲）。

Starting from the principle of creating a space of personating and living, we divided the East-West giant workshops into small ones for exhibitions of the Shanghai Expo Park, by the method of curtailment in size and intertexture in function; the original divided urban space in North-South direction by historical giant workshops is got through to organize a newly built living building——the urban complexity.

城市经纬

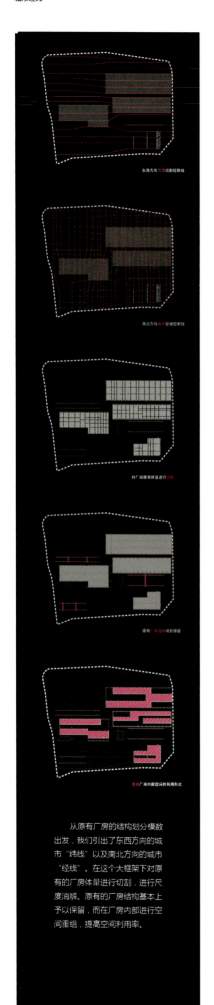

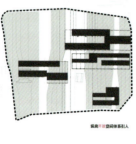

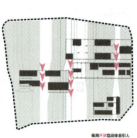

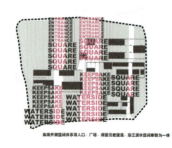

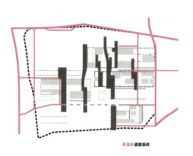

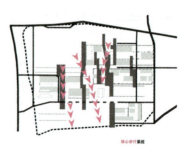

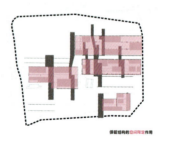

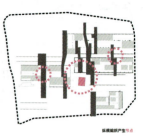

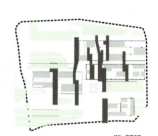

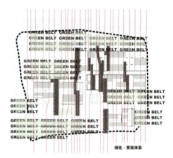

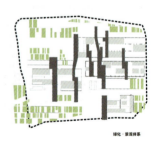

## 分析 ANALYSIS

从原有厂房的结构划分模数出发，我们引出了东西方向的城市"纬线"以及南北方向的城市"经线"。在这个大框架下对原有的厂房体量进行切割，进行尺度消解。原有的厂房结构基本上予以保留，而在厂房内部进行空间重组，提高空间利用率。

改造后的厂房部分将是主要的博物、展览区域。在改造厂房的基础上，我们引入纵向开放空间体系，打通南北方向由入口广场到沿江滨水空间的通道。在这几条通道上串联了一系列广场、纪念性建筑，成为场地内人流量最大，最具活力的区域。

结合这些贯通南北的开放空间，我们引入了纵向的建筑体块——城市活力综合体。新加入的体系与原有建筑体系垂直，由城市穿插进入厂房，指向江岸，把整个地段内的厂房紧密连接到一个体系之中。功能上，城市综合体是对世博会博览建筑的很好补充，通过多样化城市功能的引入激发整个区域的城市活力。

场地中绿化体系呈带状分布，延续建筑的大走势。在绿带之中通过经纬线的切割，划分出小块的绿地景观系统和人行道路铺装。在绿地之中结合原有的工业构筑物，形成具有地段特色的室外展示区和主题公园。

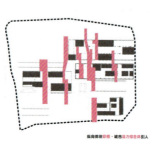

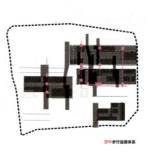

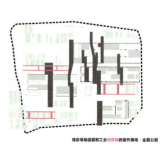

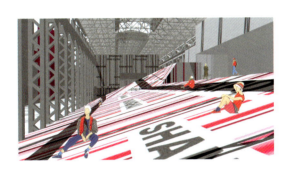

# RECONSTRUCT 厂房改造

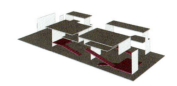

　　地段东侧的两个大厂房——船体联合仓库、西区装焊车间也是场地内主要的展览场所。

　　这一区域主要分为4个展区。西北角为铁道、火车主题展区，东北角为生态技术馆，中间是航空航天主题馆，南侧是一片室外工业遗产展区。从剖面示意可以看出，原有的工厂厂房整体大空间在结构框架的限定之下进行空间整合与重组。火车主题馆部分利用原有三跨结构的中间跨形成一个半室外的展陈空间，而航空主题馆则利用旧厂房的高空间和原有桁车结构层形成了一层空中高架展览系统。

部件装配车间设计为汽车展览馆。设计时根据尺寸划分的原则将原有的大尺度厂房保留结构,并在内部新建小尺度的展示馆。展览馆分布在行架限定的大空间两层,中间留空作为通行的空间.在交通及功能的组织上,依据汽车展览馆的展示需求,将人行和展示车辆行进的流线进行交叉设计,展览馆设计在人行线路沿线,而车辆通道在空间行相互交流却不交接.

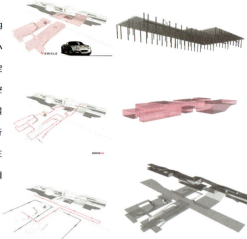

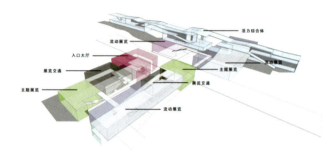

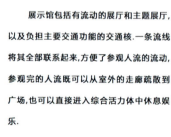

## 厂房改造 RECONSTRUCT

展示馆包括有流动的展厅和主题展厅,以及负担主要交通功能的交通核.一条流线将其全部联系起来,方便了参观人流的流动,参观完的人流既可以从室外的走廊疏散到广场,也可以直接进入综合活力体中休息娱乐.

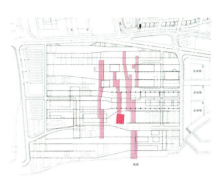

# URBAN COMPLEXITY 城市综合体

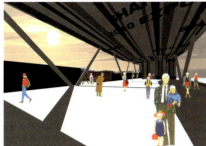

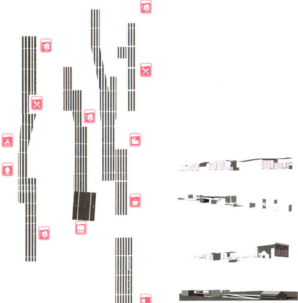

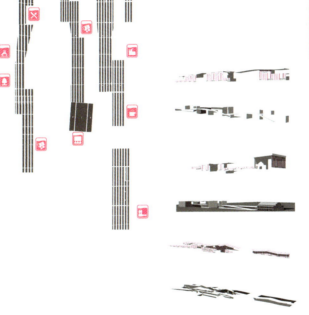

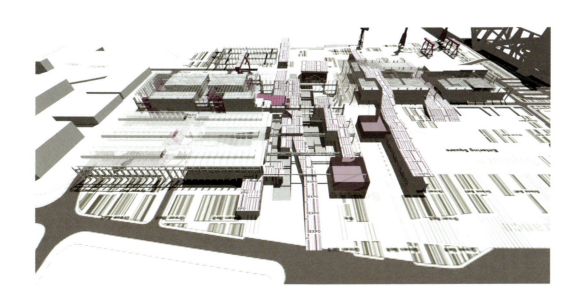

在南北方向引入城市活力综合体的概念，在与原有肌理垂直的方向进行穿插和渗透。城市综合体主体有生活性建筑和景观性通道两种主要形式。生活性建筑包括商业，娱乐，餐饮，休闲等功能，主要目的是解决世博会期间对辅助性功能建筑的需求，建立生活型博展区域，同时便于会后场地向生活型场所的转变；景观性通道设计成坡道，地景通道，空中连接平台等，主要目的在于加强场地南北向连接，打通原先由于巨型厂房的存在而被割裂开的城市空间。地景通道也可以与两侧展览空间结合，增加连接处的空间趣味和对游客的吸引力。

中心主要综合体结合飞机库设计成生活型广场，同时改造飞机库内部功能，改造为露天剧场，增加场所氛围。同时强化外立面处理，强调对历史建筑的态度。

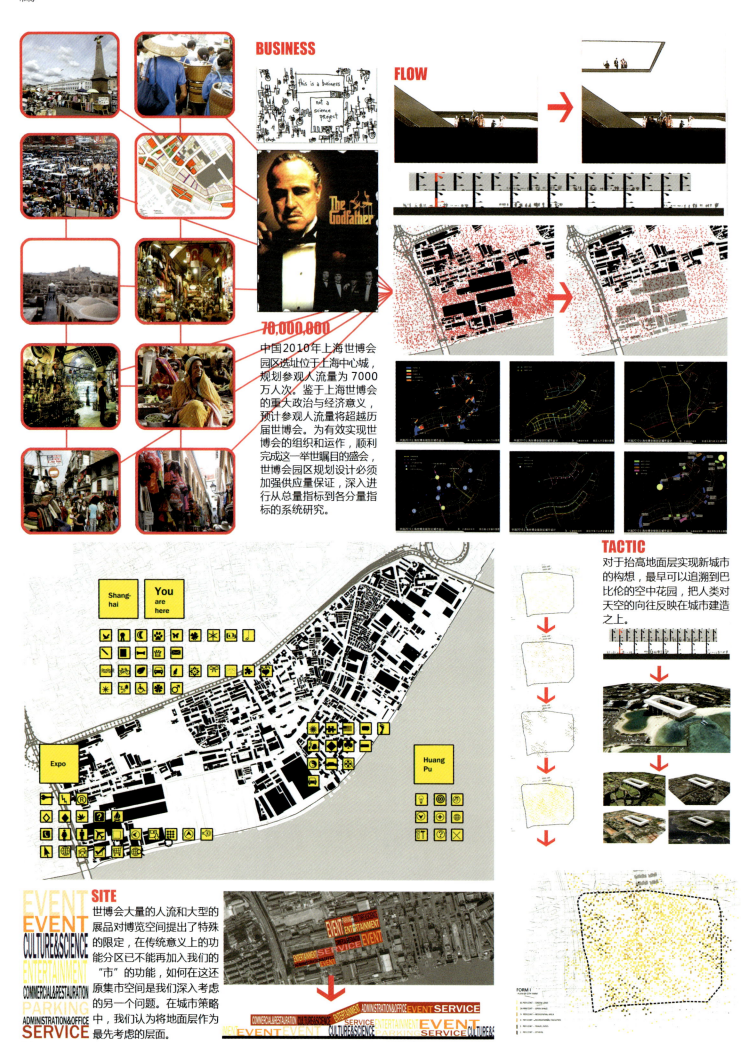

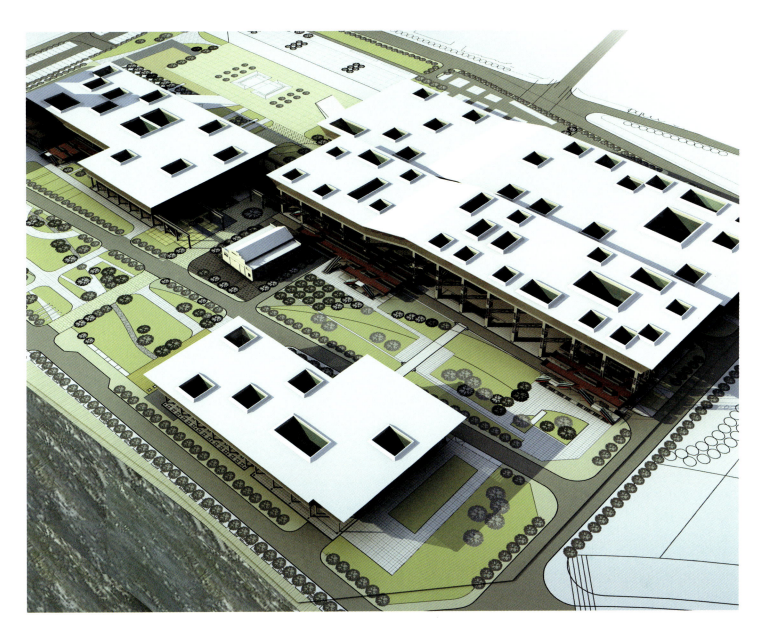

清华大学 〉市·博 〉设计：梁其伟 李娜 郑瑄 戴南
Tsinghua University 〉Commershow

# 市·博 Commershow

基于架空城市观点对商业及博览间平衡性的探讨 Research in the balance between commerce and exhibition based on the view of Aerial City

本方案立足于还原世博会的本源，以及解决地段庞大的人流问题。
"市"是世博会发源的初始形态，因此还原"市"而造成与"博"之间空间的对比关系是本方案的原点。"市"的混乱，零碎，与"博"的宏大，整体之间存在一种矛盾，这种矛盾促使我们必须寻找一种新的空间形式来化解。另外，上海市博会的地段，人流庞大，交通复杂，对地面的需求相当突出。我们因此探讨了一种最为理想的解决方式——开放整个地面以容纳这种人流量，并使得建筑脱离地面成为架空的存在，不再对城市交通造成干扰。架空体系成为我们整个设计的主体，这套解决了城市的体系又天生地形成了两个空间——稠密的空中建筑部分以及开阔的地面，正好与"市"与"博"的空间——对应。
我们将旧工业建筑作为难能可贵的支撑体系，成为架空结构的基础。原规划中完全保留的两个旧建筑成为了场地中唯一的落地建筑而成为了"博"的展品。"市"建立在原有建筑的结构之上，最大限度地利用了工业建筑的结构和跨度，开敞的"博"的空间也充分地还原了旧建筑原有的场所精神，所不同的是，这个空间现在具有了完全的开放性和通达性。

The design focuses on retrace the origin of EXPO, and resolve the large traffic flow in the site.
"Commerce" is what the EXPO began, so finding the space of "commerce" and comparing it to the traditional thinking as "show" is the origin of the design. Chaos of commerce and integrity of show cause ambivalence, calling for a new space type. In another way, the site in Shanghai EXPO is complex for its huge traffic problem with the most visitors in EXPO history, relying too much on the open space. We develop an idealized solution, opening all the ground and holding the buildings as aerial ones, no longer interacted with the on-ground traffic. As the main point of our design, the system of aerial buildings forms two spaces naturally – solid aerial buildings and void space near the ground, corresponding to the "Commerce-show" spaces.
The old structure of the original factory is identified as the basement of the aerial system. The two historical buildings that the master plan indicated are the only landing buildings, as two "showed" items. "Commerce" is on the shoulder of the old infrastructure, making the best use of the firm structure of industrial buildings, and the "Show" part rebuild the place of site, to a unique of complete open and accessibility.

市·博

"现在我要讲的城是珍诺比亚,它的妙处是:虽然位于干燥地带,房屋用竹子和锌片盖成,不同高度的支架撑住许多纵横交错的亭子和露整个城却建立于高 脚桩柱之上台,相互之间以梯子和悬空的过道相连,最高处是锥形屋顶的瞭望台、贮水桶、风向标、突出的滑车,还有钓鱼竿,还有吊钩。"

——伊塔罗·卡尔维诺 《看不见的城市》

场馆主透视

东立面图

南立面图

整个体系被清晰地划分为了三个层次，即"市"的建筑层次，"博"的场地层次。以及联结两个层次的中间通行走道。这三个层次清晰地排列在纵向高度上，自成系统，并分别与原有建筑的结构产生了不同的关系。

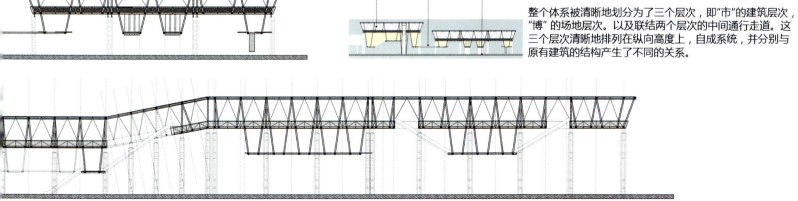

市.博

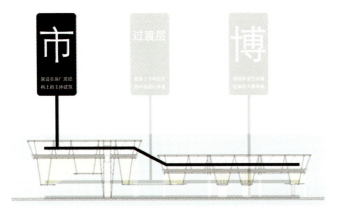

"市"是本体系中主要的建筑实体部分,也是空中城市的主体。作为还原世博会原旨的空间形式,其有意要成为并不纯粹和统一的建筑空间。在空间结构的控制之下,建筑内部单元出现了似是而非的多种分化,而基本使用单元可以被业主企业用于自用,出租,其功能也可以是商业和博览之间互换。所有的不同于集权化建筑的管理,都是让这个单体建筑更像是一个空中的城市,而不仅仅是凌空而建的又一座博物馆。

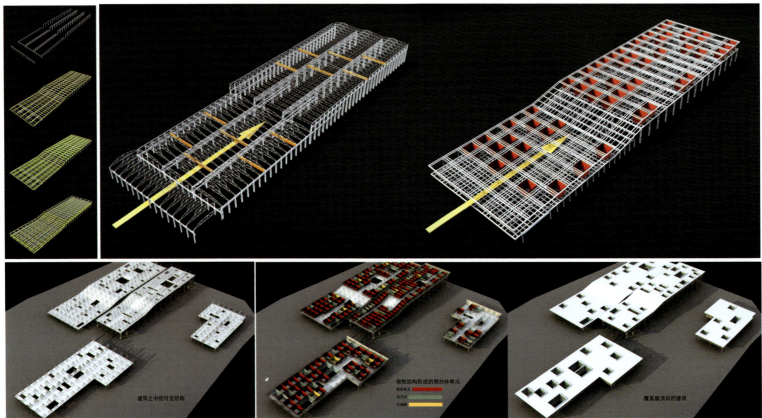

**结构体系**
在柱子顶端建立空间结构体系,对于厂房来说这样的体系是一个屋顶,对于新建筑来说,这个空间体系则是建筑单元依附的基础。

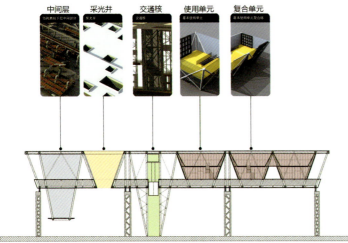

建筑结构所依附的各种功能体

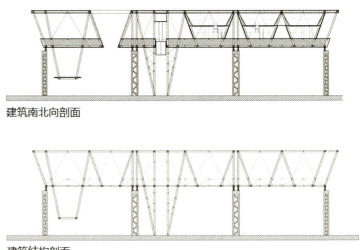

建筑南北向剖面

建筑结构剖面

街道透视

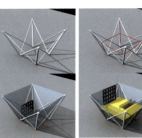
单元的形成

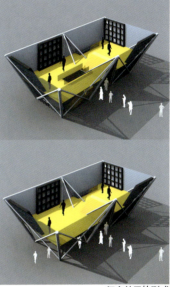
复合单元的形成

单元的使用模式
企业馆的单个场馆归某企业所有；
场馆内的单元可用于展览或商业；
企业可自用场馆单元，也可出租单元。

自用　出租
展览　商业

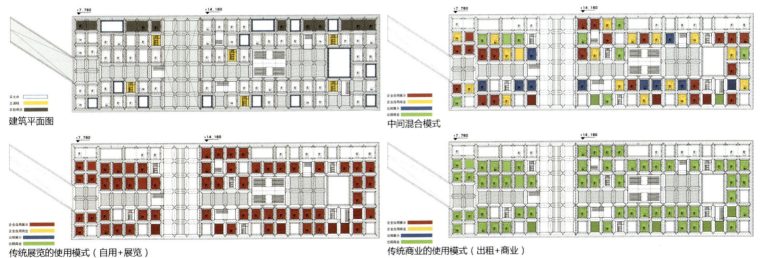

建筑平面图

中间混合模式

传统展览的使用模式（自用+展览）

传统商业的使用模式（出租+商业）

单元之内的街道形成

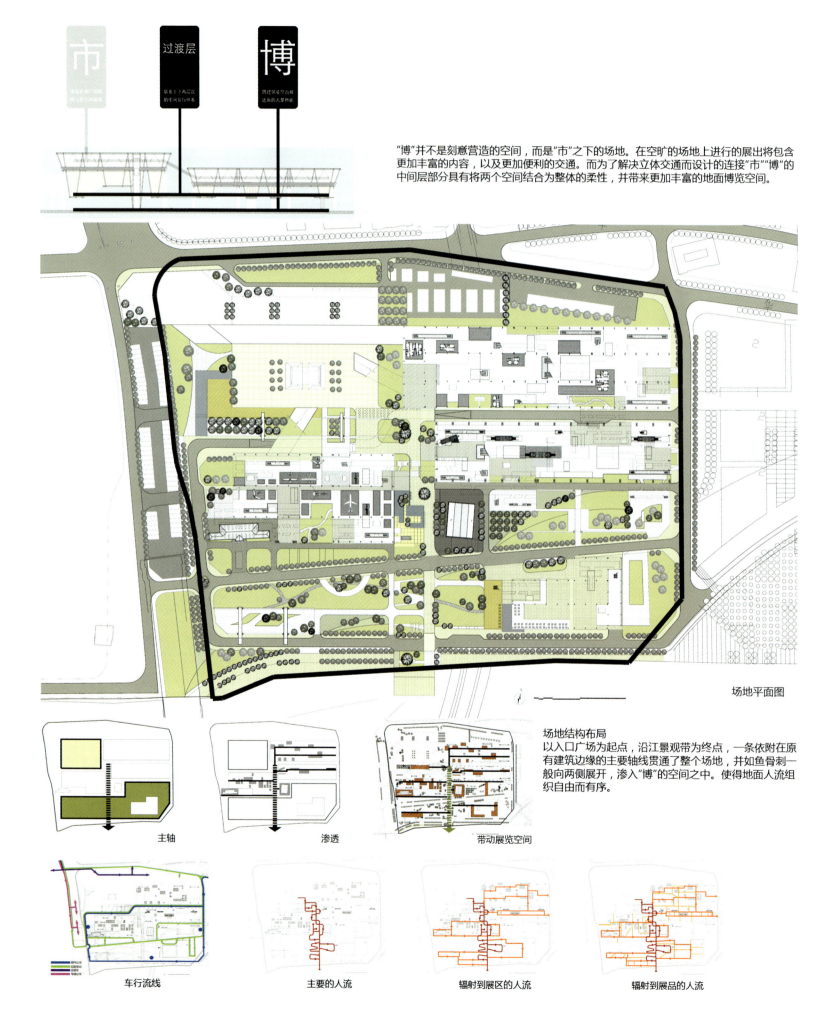

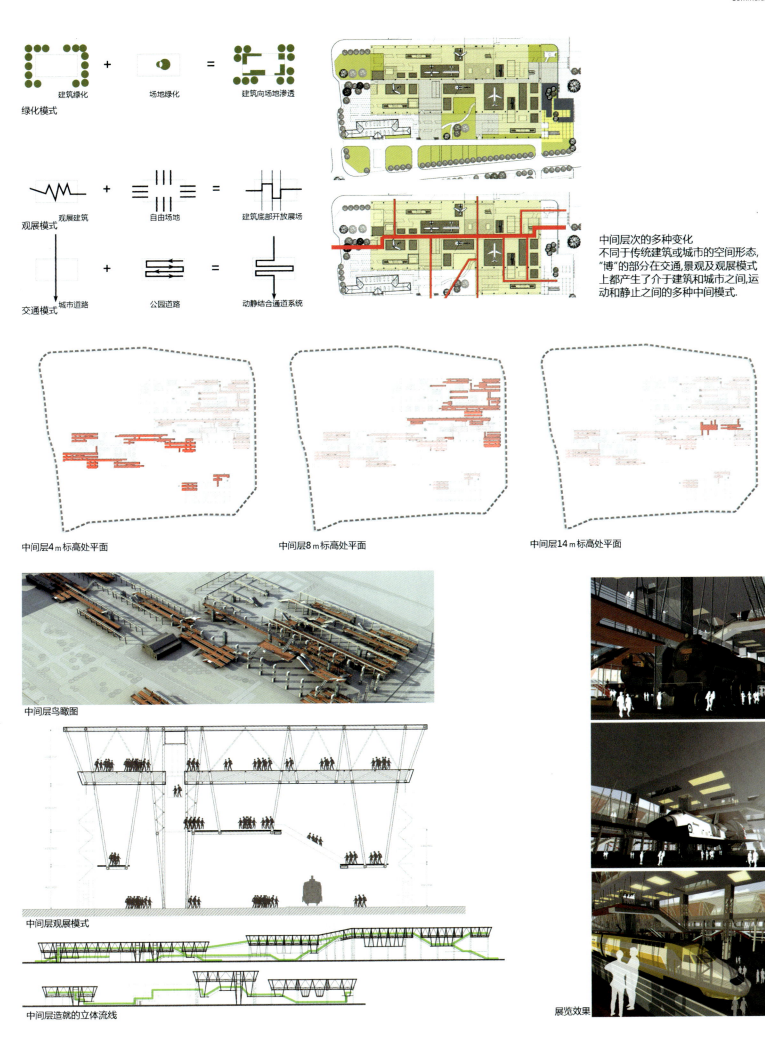

| | | | | |
|---|---|---|---|---|
| 肖 冰 | 郭东海 | 柴金戈 | 胡晨希 | 黄轩 |
| 张志兵 | 卢梨梦 | 刘彦珉 | 李瑞鹏 | 唐颖立 |
| 邱柏霖 | 刘默琦 | 申苏蓉 | 邱珏 | 崔赟 |
| 袁懿 | 李邦建 | 江天 | 何晋 | 姚昕悦 |
| 指导教师 | 王建国 | 仲德崑 | | 龚恺 |

东南大学

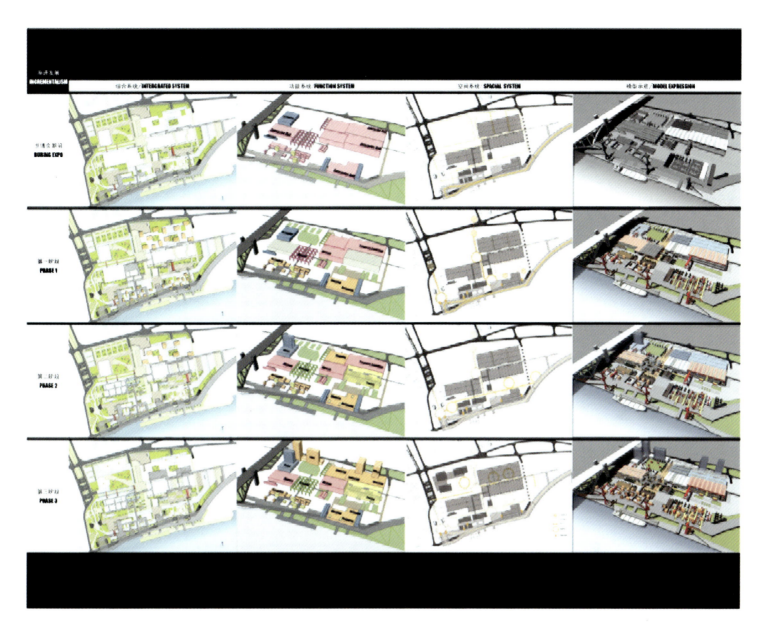

Southeast University > The Historical (Region and) Architecture Preservation and Revitalization of Jiangnan Shipyard in Shanghai

# 旧工厂适应性再生 Adaptive Revitalization of Old Factory

上海世博会中国现代工业博物馆群设计 National Modern Industrial Museums Complex Design After Shanghai EXPO

在经历了大工业时代的繁荣后，那些曾经位于城市边缘的工业厂区逐渐退化，成为不断扩张版图的新城市中衰竭的器官。上海江南造船厂就是这样一个例子。世博会的选址无疑给这一地区的发展提出了全新的命题，自上而下的力量旨在彻底改变其乃至黄浦江沿岸的面貌。船厂搬迁，世博布展，空前人潮，盛极一时。然而一剂强心剂并不能永久维持，世博结束，喧闹归为平静，人们的热情逐渐冷却，船厂故地面临萧条以至被遗忘的境地。

那么究竟怎么能够继续保持吸引力？如同生命体的生长、进化一样，江南造船厂也是个有机体，只有不断的改变自身，才能适应环境需要。所以我们从变化的角度来看待这片基地，它没有固定不变的形态规划，而是依据我们提出的渐进式策略不断改变和调整，最终进化为一个以工业博物馆为主题、多种公共及商业活动辅助的城市新中心。此为变化的时间层面。但这个稳态也并不是一成不变的，不同时节及特殊事件会对场所有不同的需求。设计方案的建筑层面即是我们的变化这一主题在空间上的探索。建筑适应其原有结构及主要功能，同时为各种性质的活动提供可能性。通过时间与空间上的变化策略，使整个江南造船厂区以新的方式复苏，并保持活力长久。

After the prosperous industry age, the industry plants which once lay at the edge of cities gradually degenerated. They became decayed organs in the constantly expanding cities. The Jiangnan Shipyard in Shanghai is a typical example of this. The location of EXPO undoubtedly put forward a completely new proposition. Top-down force aims to change thoroughly this area and even the river shore of Huangpu. Factories moving out and EXPO moving into, unprecedented number of people will make this place ever prosperous. However, a dose of cardio tonic cannot maintain its vigor. After EXPO, people will cool down and the shipyard will have to face depression again.
Then how to maintain activity continuously? As a growing living thing, The Jiangnan Shipyard is also an organism. It can adapt to the environment by changing itself. Therefore we look on the area at a CHANGE point. There is no fixed planning. It will change and adjust basing our incrementalism strategy, and finally become a new center, of which industry museum is the main theme, accompanying various public and commercial activities. This is the time aspect of our concept CHANGE. This steady state also changes. The demands of different seasons and special events are different. Our building design is to study the space aspect of CHANGE. The buildings and site provide possibilities for various kinds of activities. Through time and space change strategy, the Jiangnan Shipyard will revive and keep prosperous in new method.

旧工厂适应性再生

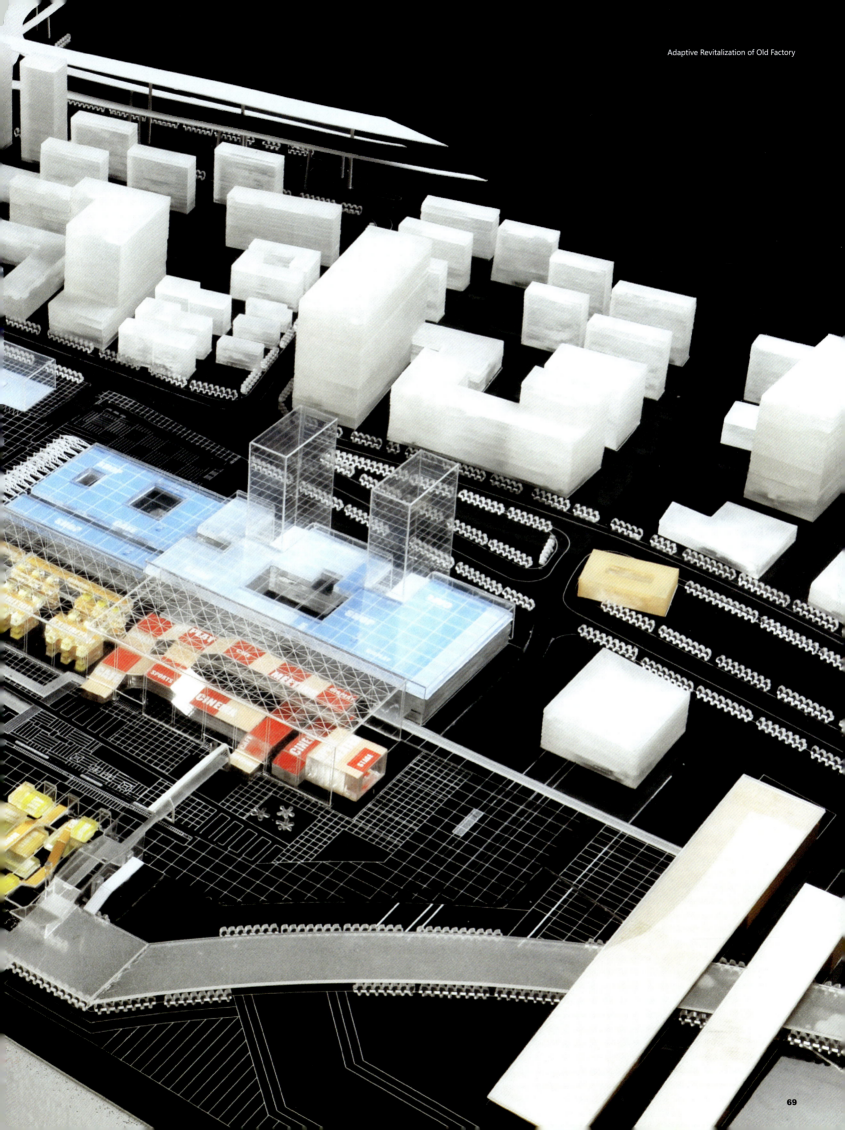

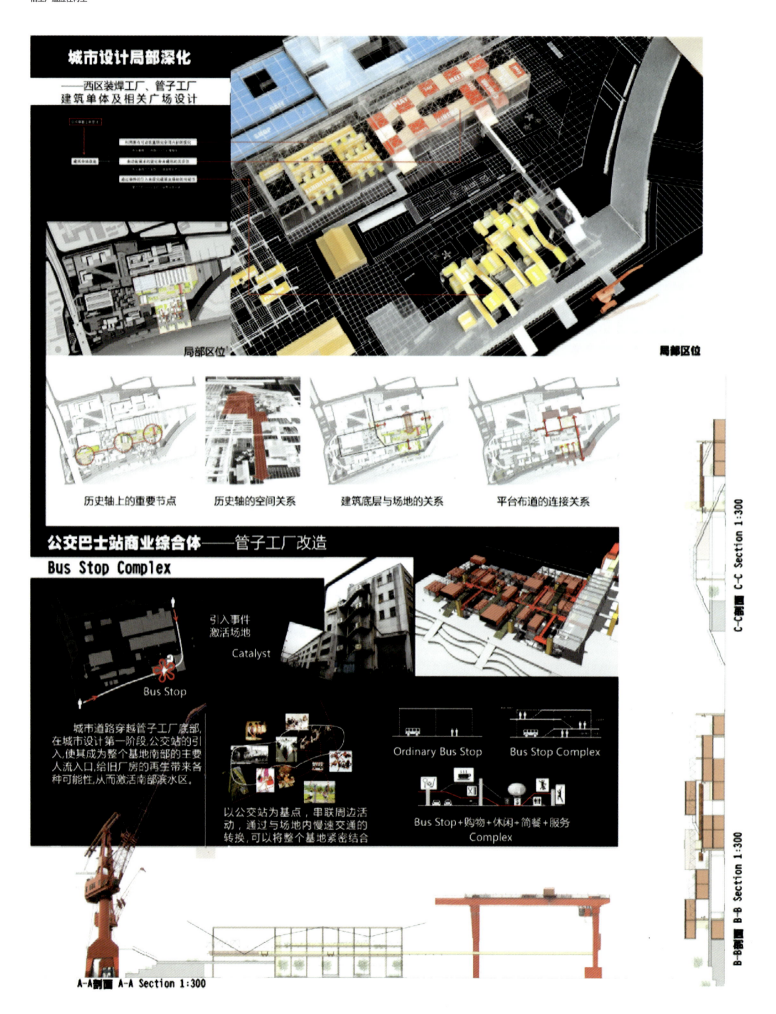

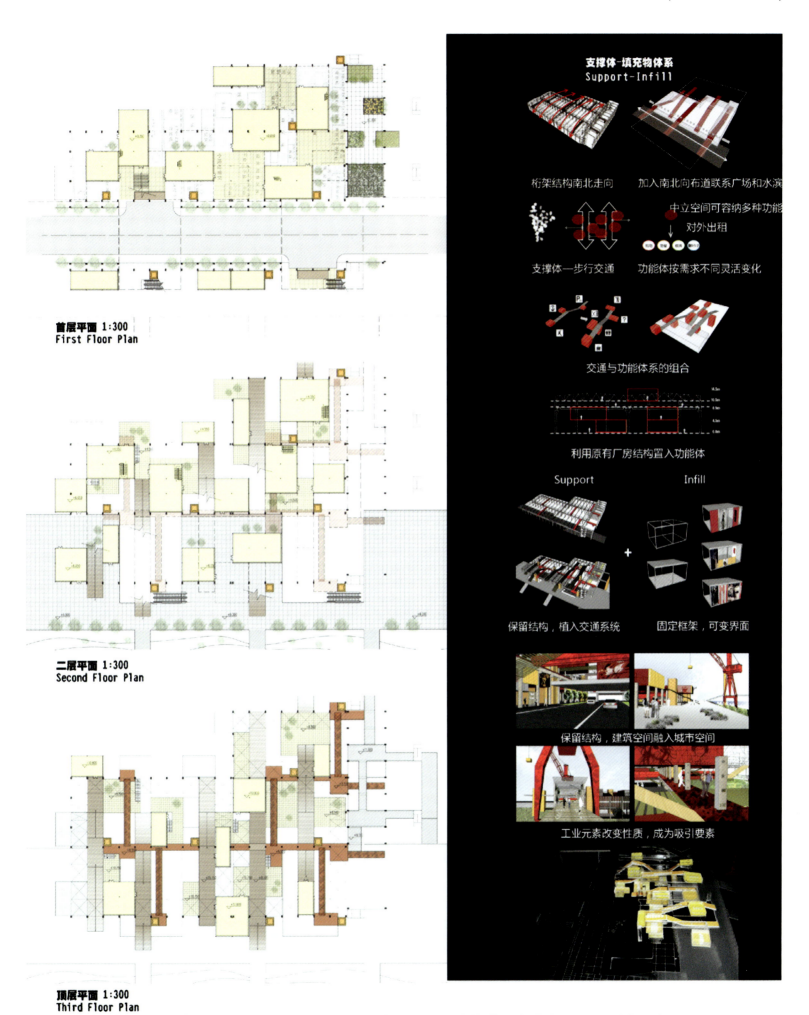

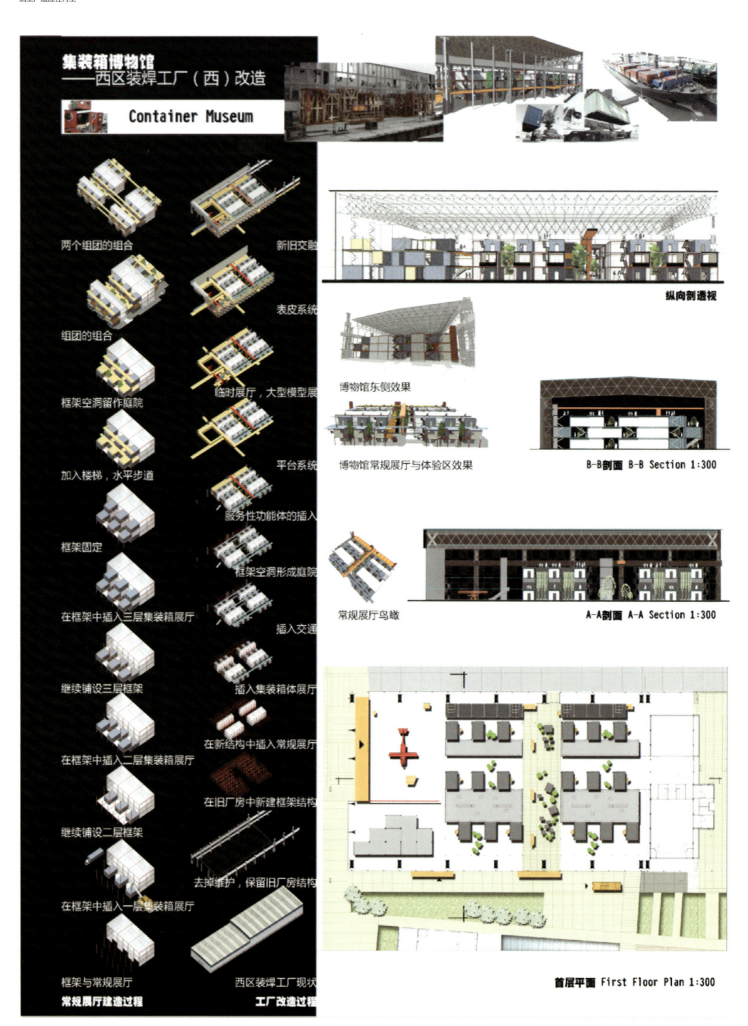

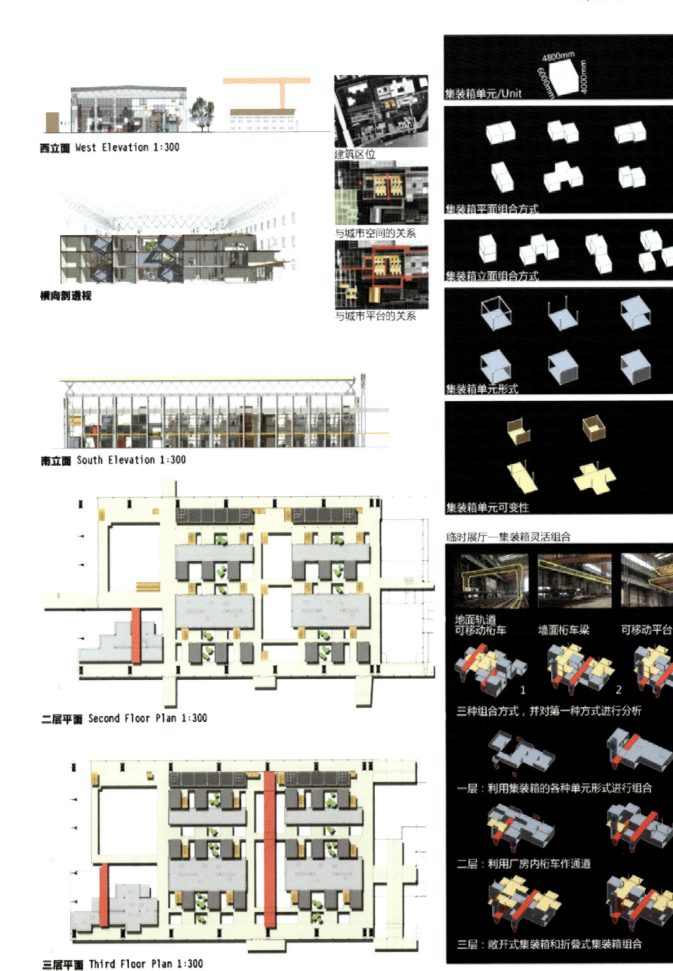

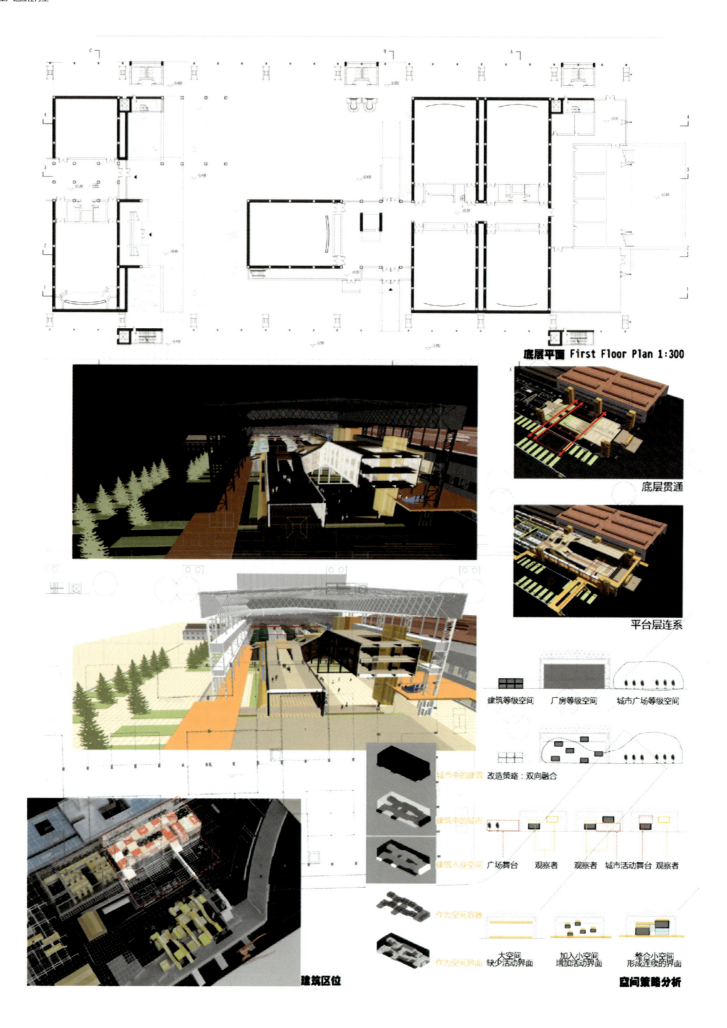

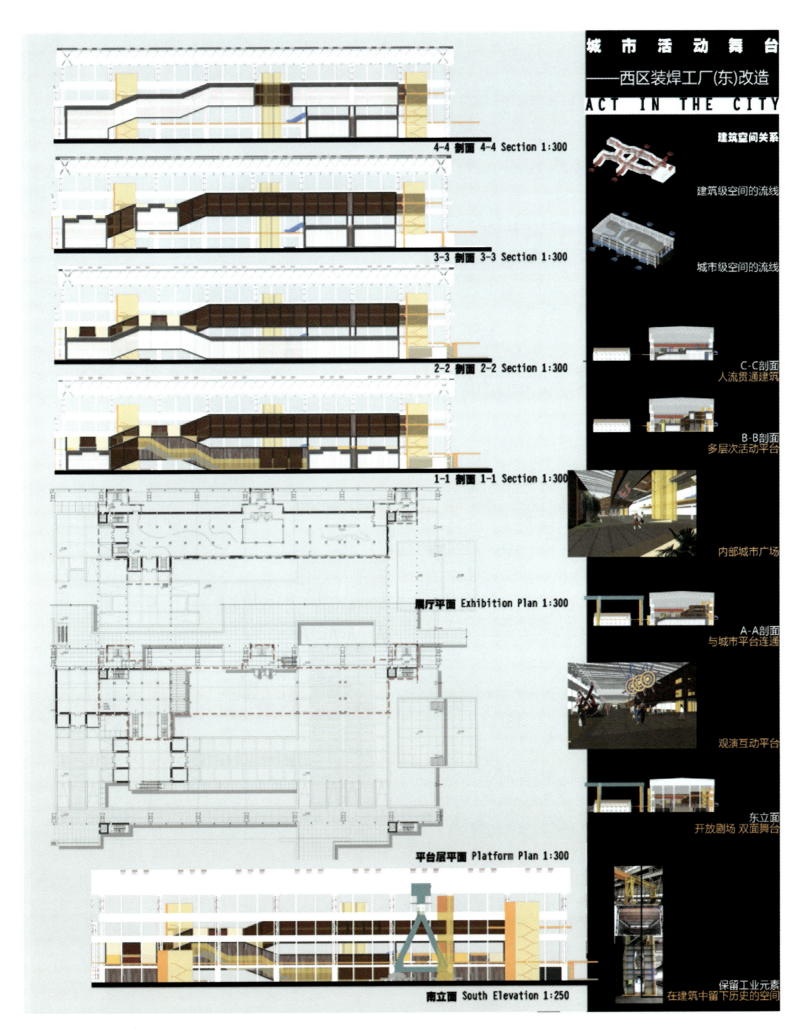

江南造船厂历史地段之肌理再生

上海的发展经历了四个时期:海滨县城,海港城市,工业基地,经济中心
社会影响下,城市结构一直处于动态的调整过程中,城市形态不连续……

城市肌理体现城市特色,是城市最直观的识别性之一

城市形态的五项评价标准:
1. 有益个体的生活环境
2. 可知觉性:可识别的、结构化的、有场所感的知觉环境
3. 形式适合行为的需要
4. 可通达性
5. 人对环境的控制

凯文·林奇《好的城市形式》

在这种纷繁芜杂的城市结构中,人对于 纹理组织、交通模式、居住密度 的感知成了认知城市的主要途径。
——凯文·林奇

城市肌理
断裂、缺失……

## 肌理分层

基地三因素　　城市建筑肌理　　工业建筑肌理　　滨水建筑肌理

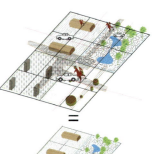

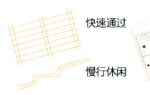

快速通过
慢行休闲

从城市中来　　滨江慢行道　　穿越工业厂房

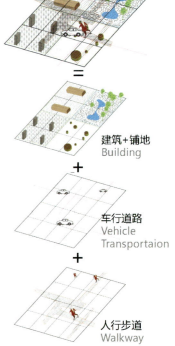

= 建筑+铺地 Building
\+ 车行道路 Vehicle Transportaion
\+ 人行步道 Walkway

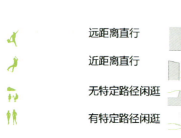

远距离直行
近距离直行
无特定路径闲逛
有特定路径闲逛

二层架空步道,串联各种空间节点,引向滨江休闲区

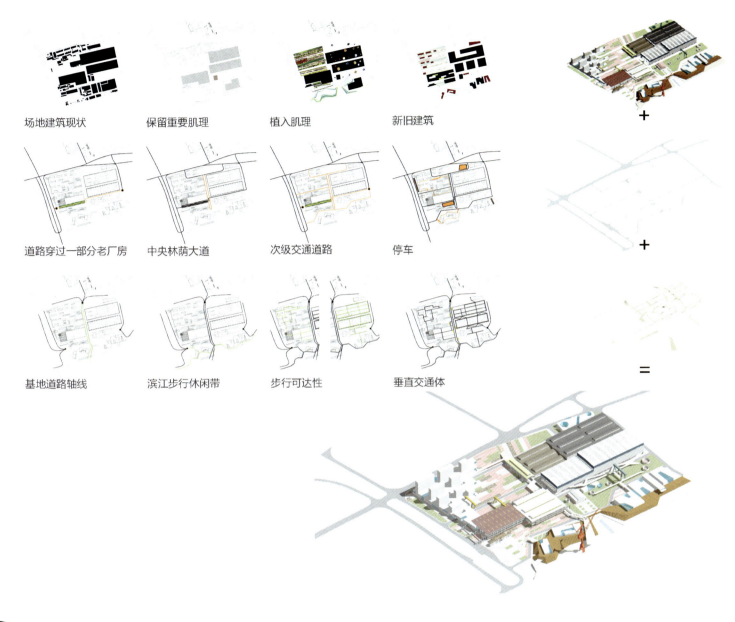

**东南大学** 〉江南造船厂历史（地段和）建筑保护和再生　设计：胡晨希 柴金戈 肖冰 郭东海 黄轩
Southeast University > Designed by HCXGH

# Urban Fabric Transplant
## 江南造船厂历史地段之肌理再生
上海世博会中国现代工业博物馆群设计

从1865年江南制造局在此创立至今，历经辉煌，见证时代变迁，如今面临被拆除的尴尬，
在2010年作为世博会企业馆区一部分后，这块历史地段又将如何更新发展？

俯瞰场地周围的城市环境，我们直观的发现城市生活区的小尺度肌理在此突然消失，形成了不和谐的断裂地带。
凯文·林奇在《好的城市形式》中给出城市形态评价的五项标准中，第二项是城市应该有可识别的、结构化的、有场所感的知觉环境。
因此我们选取肌理作为我们城市设计的方向，重生江南造船厂地带的城市肌理，使该地从工业向城市转换。
在这种纷繁芜杂的城市结构中，人对于纹理组织、交通模式、居住密度的感知成了认知城市的主要途径。
在此理论基础上，抽取城市最基本的元素建筑、车道、人行，强化每一元素形成的独特肌理，在将他们叠合形成抽象城市。
基地南侧临水，拆除破旧建筑植入新建筑，处理尺度巨大的工业厂房构成该城市设计的主要问题，建筑的肌理重点考虑城滨水建筑与水的渗透关系、
新旧建筑之间构成的拼贴关系和尺度之间的对比。

Since the Jiangnan shipyard was funded in this site in 1865, it gained glories and fames, witenessed the change of epochs, and now faced the embarrassment of being teared down. After as a part of the Corporate Pavilions in World EXPO 2010, where will this historical site goes ?Through overlooking it's circumstance, we can easily tell that the small living scale was suddenly disppeared here and formed a blank of city grid.
Among five standards to estimate city forms raised by Kevin Lynch in his 《Good city form》, the second peice is, city should be provied with characteristic, structuralism and sense of place in perceptable enviroment.
So we choose the texture research as the theme of urban plan, try to rebuild the urban-fabric here, for it's transformation from a brown-field to an public space.
Perception to the fabric, transportation and density build up the main approach to the city. So we select the essential elements - buildings, roads and walkway, to intensify their unique.

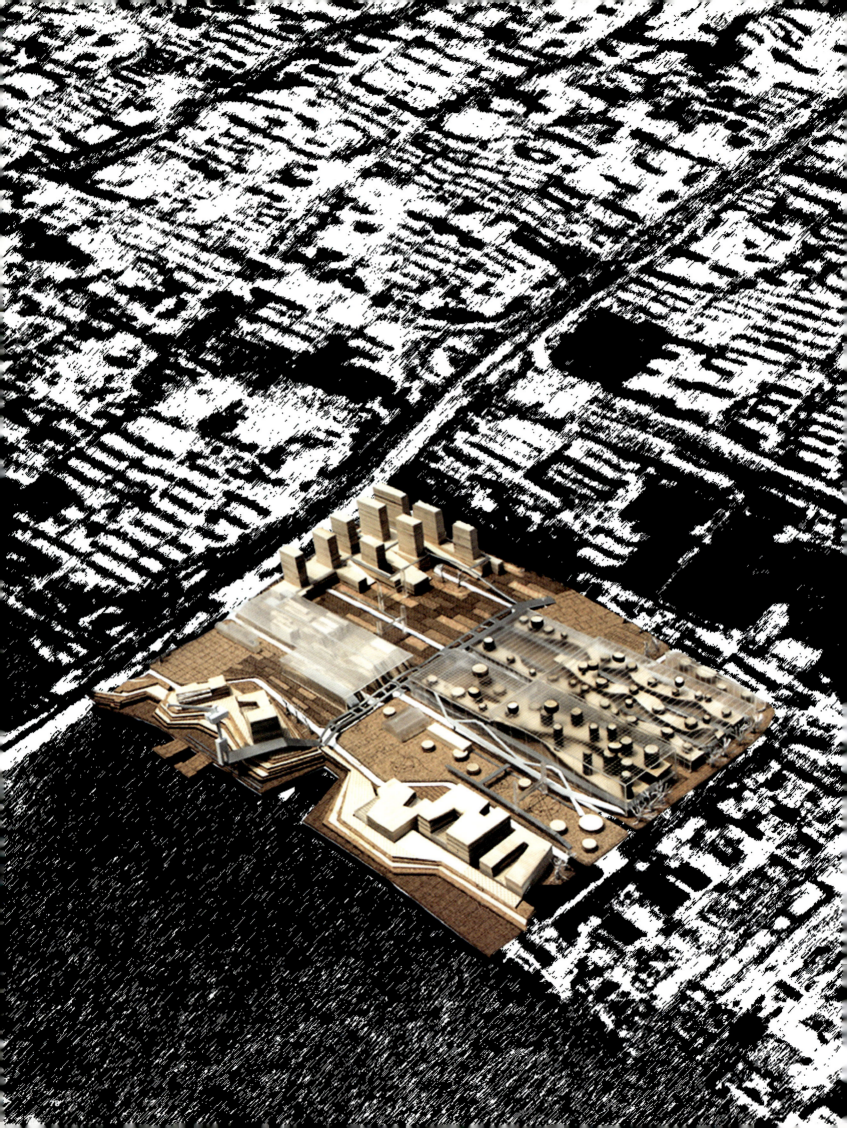

总平面

经济技术指标：
总建筑面积 Floor Area　　　330,000m²
容积率 Floor Area Ratio　　　1.4
建筑密度 Building Coverage　38%
绿地率 Greening Rate　　　　32%
停车 Parking：　　　　　　2000（地下）200（地上）

江南造船厂历史地段之肌理再生

## 建筑肌理研究

### 水岸渗透

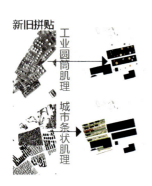

设计堤岸标高：6.0m

## 业态分布

### 滨江休闲区

娱乐
滨江旅道
餐饮
起重机
滨江旅馆

### 新旧拼贴

工业圆筒肌理
城市条状肌理

### 工业博物馆群

工业体验中心
造船博物馆
历史博物馆
飞机展览馆

### 大小对比

西区装焊工厂改造
保持体量
插入小尺度形体
强化对比

管子工厂改造
化整为零
打碎体量
削弱对比

船舶联合车间改造
大厂房虚化体量，
插入较大尺度形体，
折衷对比

### 市民活动区

商住 购物中心
装潢工坊
工作室
市民活动场
工业展览馆

## 开放空间

A 中央步行道

B 保护建筑长廊
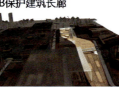

C 入口广场

D 市民活动广场

E 飞机库前广场

F 海军司令部前广场

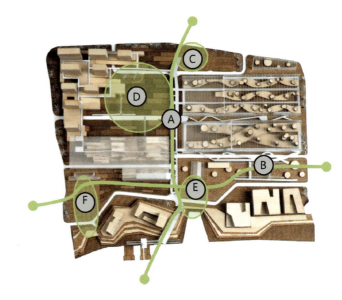

80

# 艺术工场

Southeast University > Designed by HCXGH

江南造船厂历史地段之肌理再生——西区装焊工厂改造

西区装焊工厂位于城市设计中条状肌理的市民活动区，北临市民广场南有历史保护建筑海军司令部，有一条城市支路穿越厂房，以及两条人行步道跨越其中。

工厂广场路径　　三种肌理形成的工业秩序　　折板分割空间属性

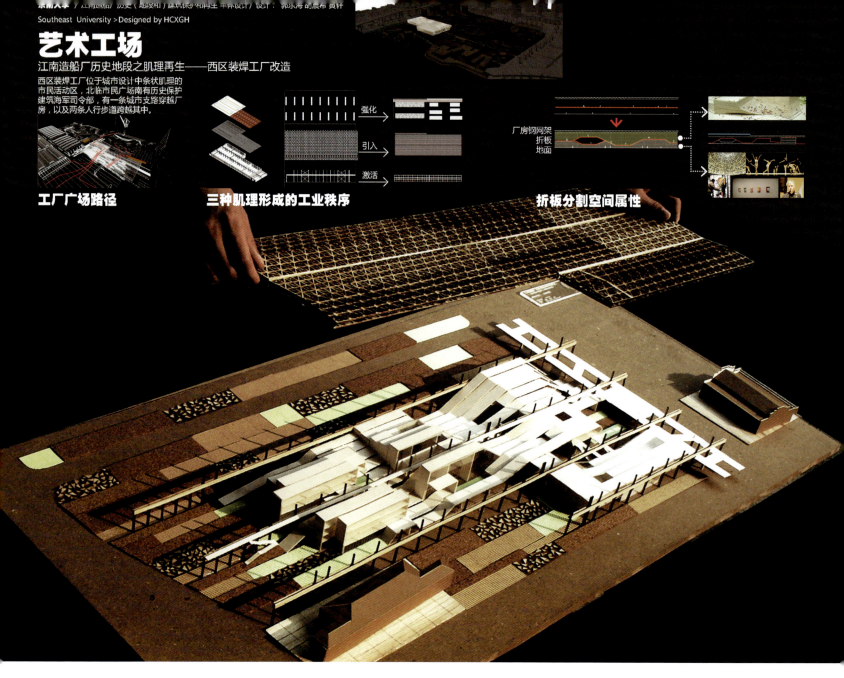

## 条形肌理类型研究

坡道　　剧场　　展览　　工作室

## 剖面分析

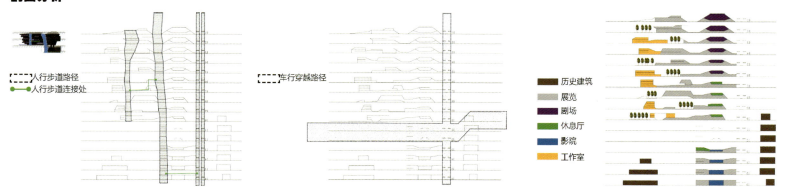

人行步道路径　人行步道连接处　车行穿越路径

历史建筑　展览　剧场　休息厅　影院　工作室

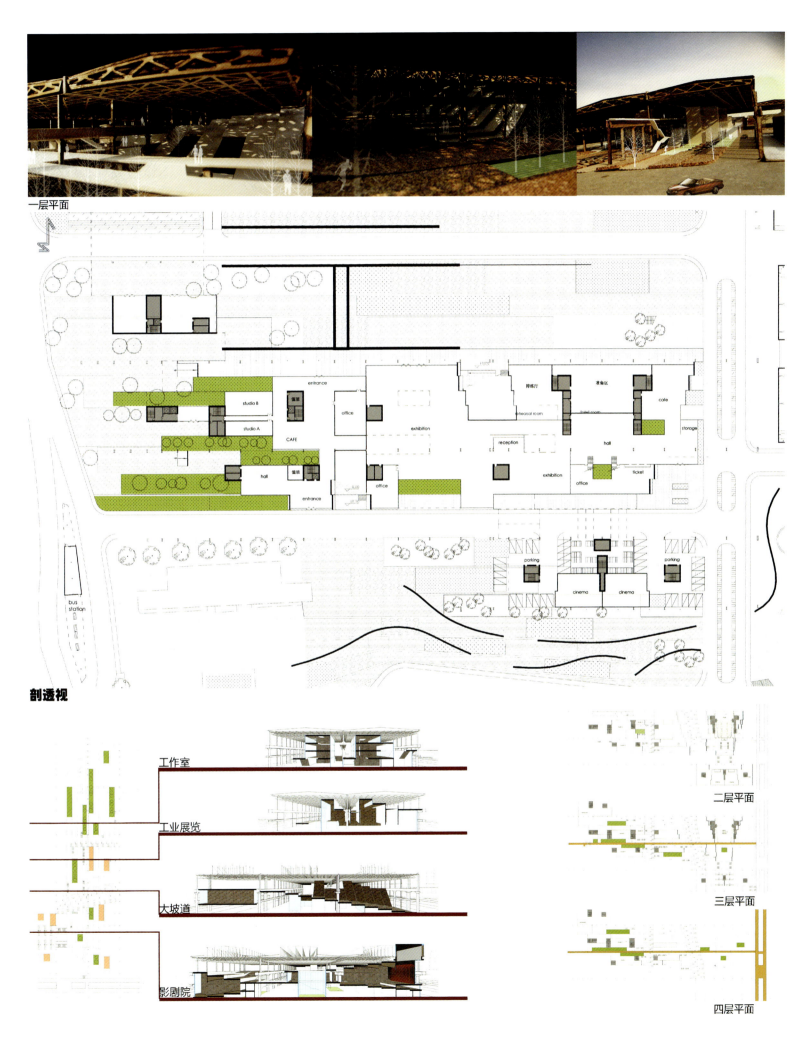

东南大学 〉江南造船厂 历史(地段和)建筑保护和再生 单体设计 设计：柴金戈 肖冰 黄轩
Southeast University > Designed by HCXGH

# "管"中博船
## 江南造船厂历史地段之肌理再生——船体联合车间改造

### 城市设计概念延续

船体联合车间位于基地东北角，在城市设计中作为工业博物馆的一部分，介入了圆筒状新肌理，作为对工业遗存的延续和敬礼。

### 造船的肌理和船的肌理

造船车间的肌理与停泊在江边的船舶的肌理都具有鲜明的特色，前者极大，后者极小；然而他们并不矛盾，他们是息息相关的。

对于人来说，造船车间是超尺度的，人在这里能感到的只有渺小。

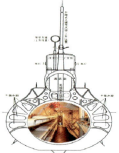
而船体空间却恰恰相反，它给人提供的是极端经济的小空间。

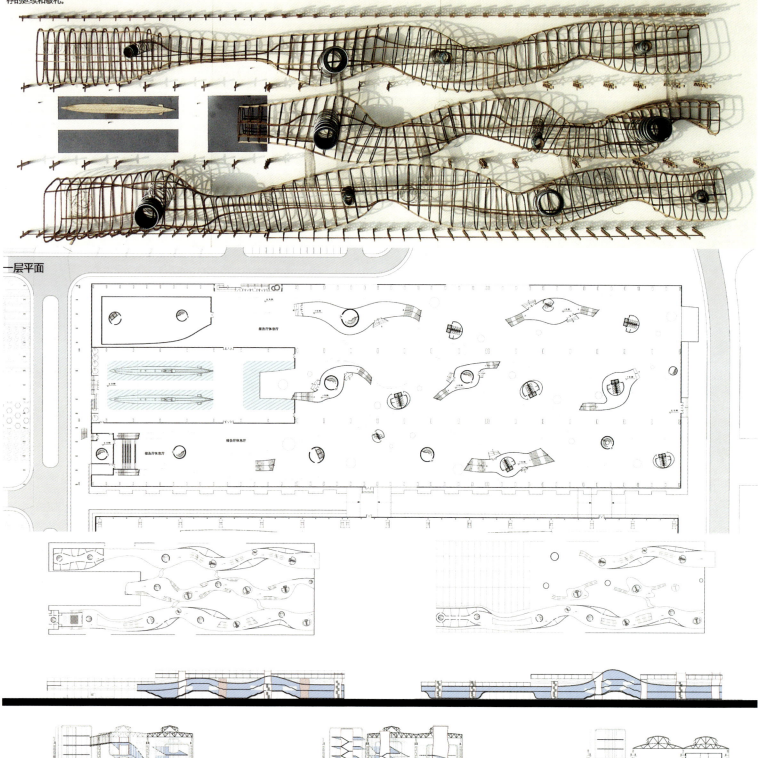

一层平面

### 管状肌理的植入

作为基地内最为典型的地块,两座大尺度的工业厂房的机理都得到了保留,并且通过调整引入了人体尺度的影响,使其与周围,尤其是市民体验去更加协调。

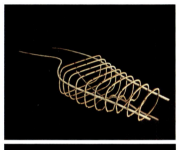
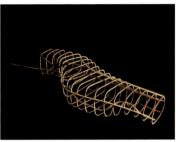
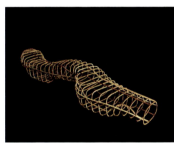
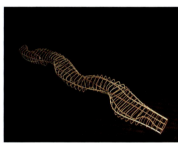

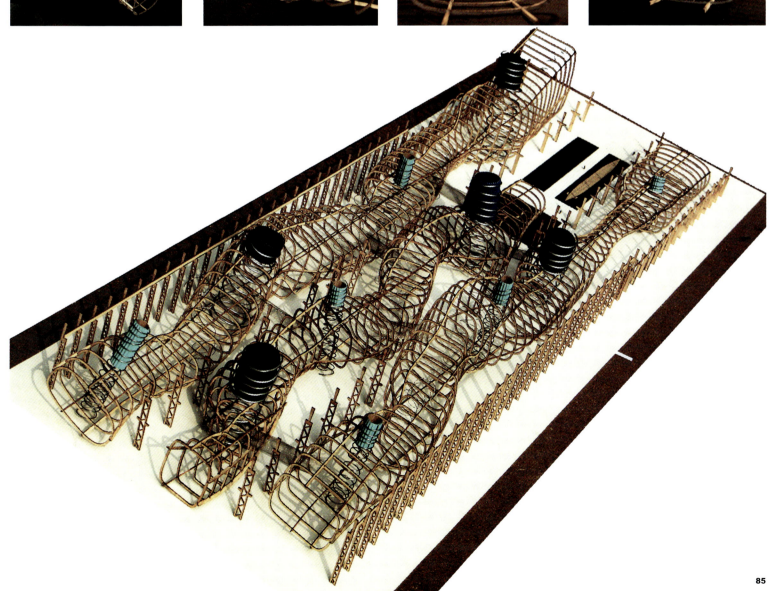

从历史禁区到城市开放空间

调研对比分析　　研究内容：在相同客观环境下改造后的开放度　　研究目的：寻求禁区到开放的操作方式　　研究方法：对比

　　　　　　　　　　　　　　　　　交通　　　　　　　　　　　　　　可识别性　　　　　　　　　　　　　功能

田子坊

8号桥

M50

中

结论：在基地周边区位交通条件相似的情况下，基地本身交通的可达性及通达性，基地的可识别性，基地内部的业态及其混杂度，将决定基地面对城市开放程度。

基地分析：禁区

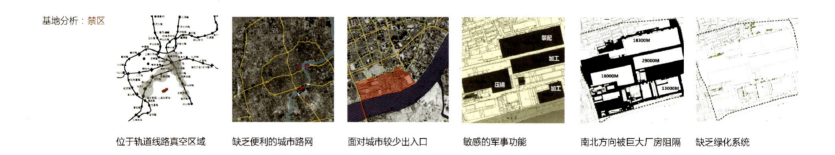

位于轨道线路真空区域　　缺乏便利的城市路网　　面对城市较少出入口　　敏感的军事功能　　南北方向被巨大厂房阻隔　　缺乏绿化系统

From Forbidden Zone to Public Space

规划用地面积：234356平方米
总建筑面积：279300平方米
地面：241300平方米
地下：38000平方米
容积率：1.2
绿地率：24%
停车：724
地面：224
地下：500

总平面　　1：1000　N

**东南大学** 〉江南造船厂历史地段和建筑的改造和再生----上海世博会中国现代工业博物馆设计 设计：申苏蓉 邱珏 崔赟 刘默琦 邱柏霖
Southeast University >Adaptive Reuse of Jiangnan Dorkyard----Shanghai Modern Industry Museum Design

# 从历史禁区到城市开放空间
## From Forbidden Zone to Public Space

一个半世纪以来，从一块城市的飞地到完全对普通市民关闭的工业基地，江南造船厂对于城市来说始终是一个禁区。随着2010上海世博会的到来，运转百年的老工厂将被迁移，这将成为这一历史地块回归城市回归生活的契机。

本方案从研究城市地块开放性的条件入手，在各方面提出了让这一地块开放并融入城市，最终成为新的城市活力点的对策。在交通方面，提出轴网，东西向自下而上尊重基地原有路网，南北向自上而下加入以步行为主的纵贯道路，以梳理城市通往滨水的肌理。同时，我们提出支撑体系的概念，在场地中加入支撑核，让城市公共功能集中，最大限度的释放公共空间使之可以更加灵活可变的生长，以满足不同业主的需求，增强场地活力，并使得更多空间回归市民。为了让地块保有持久的活力，我们在以主题博物馆为基地主导功能的条件下，分析基地周围条件以及其本身气质，进行功能策划，一方面满足不同人群的需求，让该地块成为一个复合型的用地，另一方面也能在经济平衡方面提供保障，给基地带来了最大的活力。原先巨大尺度的空间先被化整为零，后又因为各功能的需求被重新自由灵活的组合，高效地服务于各类人群，以开放的姿态融入城市，并将与城市共同得到更好的发展。

This plan from studies the urban land parcel open condition to obtain, proposed in various aspects lets this land parcel open and integrate the city, ends in a new urban vigor countermeasure. In the transportation aspect, proposes the axis net, the thing to respects the base original road network from bottom to top, the north and south to joins from the top downward walks the north-south path primarily, combs the urban path shore water the skin texture. At the same time, we proposed that the support system's concept, joins the support nucleus in the location, lets the urban public function be centralized, the release public space maximum limit enables it to be possible a more nimble invariable growth, meets the different owner's need, strengthens the location vigor, and causes more spaces to return the resident. In order to let the land parcel hold the lasting vigor, we in take the subject museum as under the base leadership function's condition, analyze the base ambient condition as well as itself makings, carries on the function plan, on the one hand satisfies the different crowd the demand, enables this land parcel to become a multi-skill the land, on the other hand can also provide the safeguard in the economic equilibrium aspect, has brought the biggest vigor to the base. The huge criterion's space breaks up the whole into parts first originally, latter because of various functions' demand by again the free nimble combination, highly effective is served each kind of crowd, integrates the city by the open posture, and will obtain a better development together with the city.

从历史禁区到城市开放空间

城市设计策划：开放

周边城市交通

建筑去留

疏通基地

交通策划

支撑体系

支撑核

业态策划

业态混杂:熵

支撑核+熵

依据场地原始框架模数建立单元空间模数，并加入次一级支撑核建立骨架。设计一套空间秩序生长规则，为活力注入创造多样可能性，加强基地开放程度

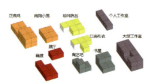

轴网　　　　　功能核分布　　　机动交通　　　步行交通

地下交通　　　电瓶车流线　　　二层步道　　　绿地景观系统

From Forbidden Zone to Public Space

南北剖面 A-A

南北剖面 B-B

南北剖面 C-C

南北剖面 D-D

空间结构

中轴线

Z字型广场

商业SOHO区

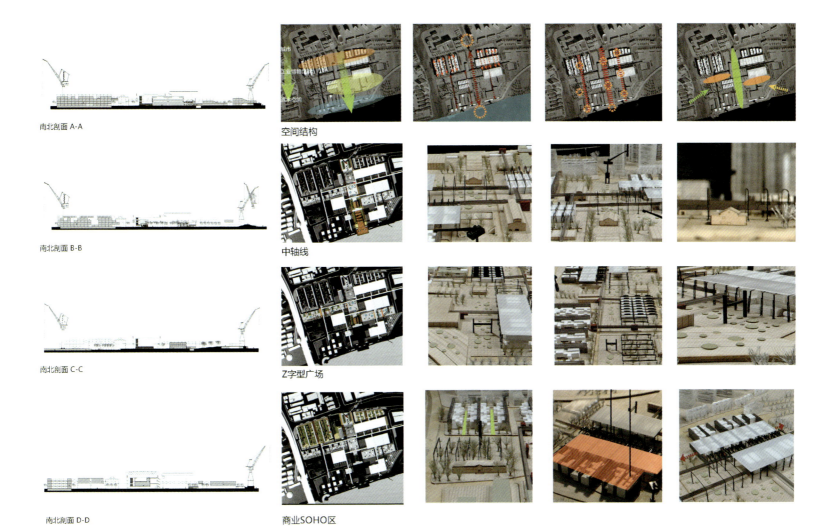

从历史禁区到城市开放空间

建筑设计——西区装焊工厂改造 > 设计 > ：申苏蓉 邱珏 崔赟

# 中国航空航天博物馆

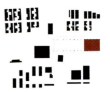

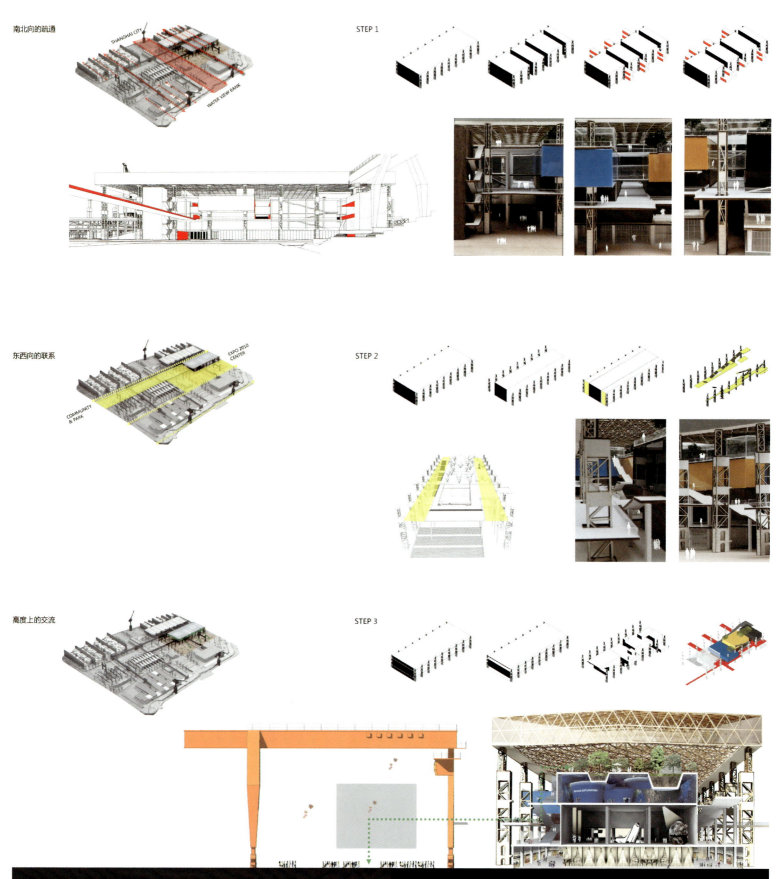

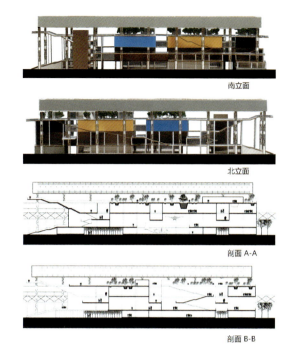

南立面

北立面

剖面 A-A

剖面 B-B

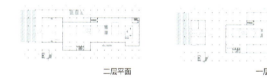

| 四层平面 | 三层平面 | 步道层平面 | 二层平面 | 一层平面 |

# 禁区·开放 旧工业之冢——部件装焊车间改造设计

从历史禁区到城市开放空间

设计：刘默琦 邱柏霖

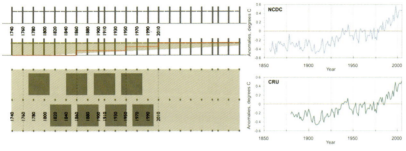

起点：歌颂工业辉煌的同时，应该铭记其残酷的历史和现实。我们设计一座工业历史的纪念碑，一座埋葬灰色记忆的家。

结构：以时间线索，分为四个部分对应三次工业革命和上海世博。内部空间用土填满，展厅是一系列洞穴，并逐渐抬升，隐喻温室效应。

空间组织：设计一条二氧化碳通道，作为异质的介入，串联展厅。

From Forbidden Zone to Public Space

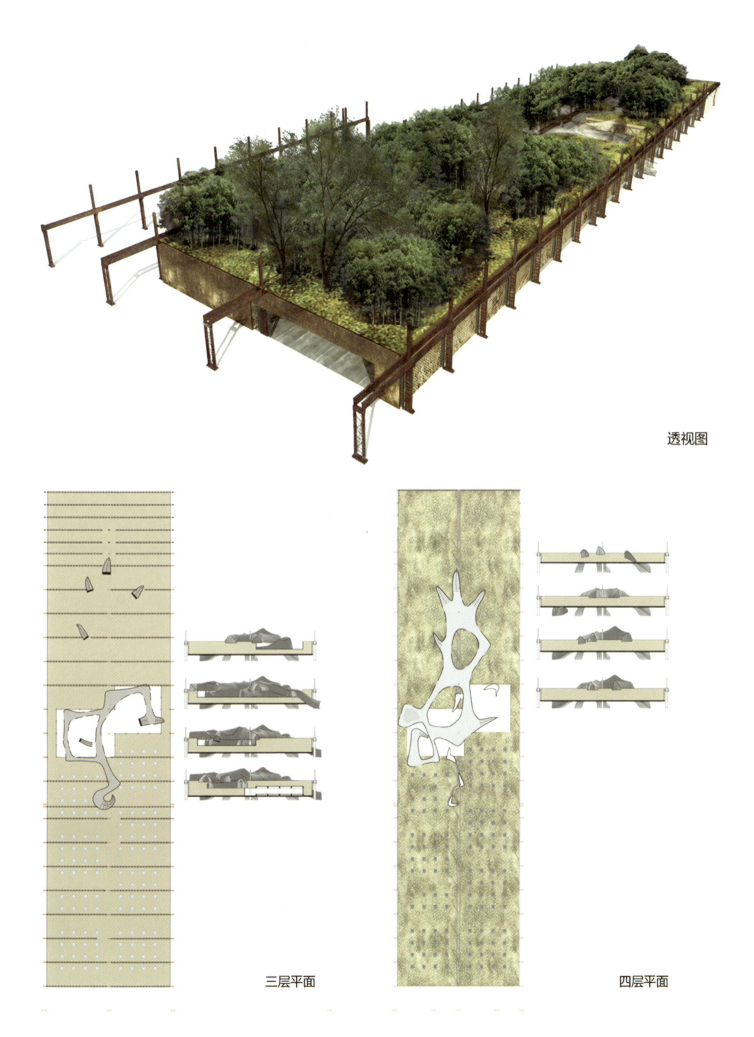

透视图

三层平面

四层平面

# 1 设计背景 Back Ground

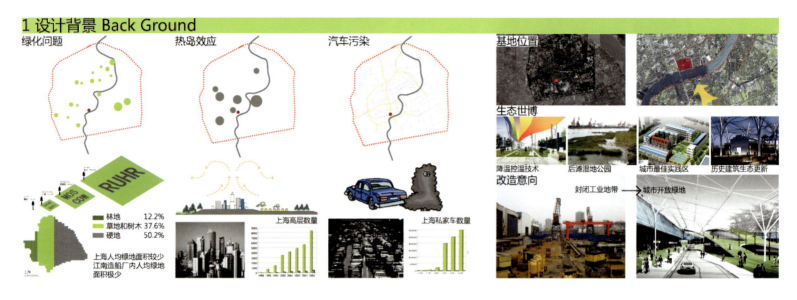

# 2 主题策划 Programme

风场策划

水场策划

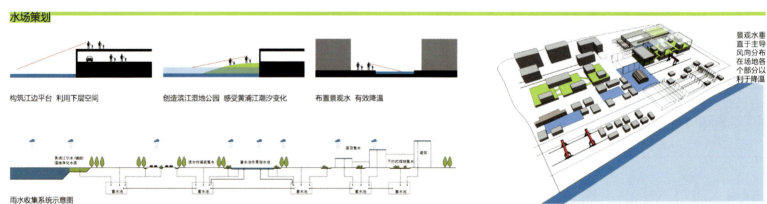

工场策划

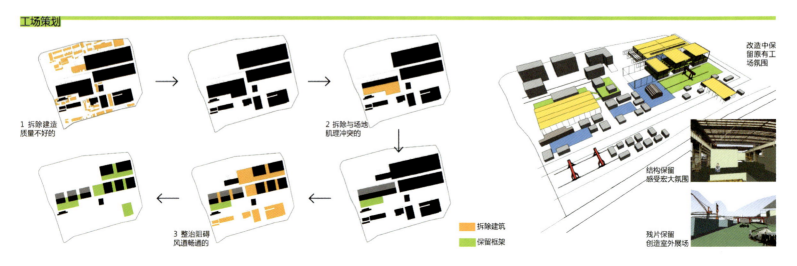

总平面图

技术经济指标
用地面积　　234,000m²
建筑占地面积　61,877m²
建筑密度　　　26.4%
建筑面积　　　277,290m²
容积率　　　　1.2

东南大学 〉风场·水场·工场〉 设计：姚昕悦 江天 何晋 袁懿 李邦健
Southeast University >Wind Water Work.Field

# 风场· 水场· 工场 WindWaterWork.Field

上海世博会江南造船厂历史地段改造和更新

近二十年来，随着上海经济社会的高速发展，城市生态环境的持续恶化日益受到关注。2010年世界博览会的举办为上海的城市生态改造提供了契机。基地位于黄浦江北岸，紧邻市中心，原为封闭的滨水工业区，改造后将成为城市公共空间。本方案由基地所处的特殊微气候环境出发，从风环境和水环境着手，并引入"生态层"的立体绿化空间，使高速、紧张的城市得以舒缓、呼吸。在生态手段上，本方案通过建筑及植被的开敞与围合，形成一系列疏风聚风的开放空间，创造场地内适宜的风环境，并通过软件计算验证逐步优化方案；充分利用水资源，通过雨水收集系统组织场地水环境，并力图达到水量平衡，在景观上活水的同时也在生态上活水，利用水体及水生植物的生态功能调节场地微气候；两条水廊同时打通了场地的历史文脉，在保留工业建筑物质结构的同时，也使百年老厂乃至上海近代民族工业在精神上得以延续；除了用以净化水源的水生植物之外，场地内还结合不同业态的文化需要、遵循风场和水场的布置原则、并且考虑对既存工业污染的治理布置树种，形成错落有致、动静有别、四季交替的景观环境。

In recent 20 years, as economics and society developed in Shanghai, the deterioration of city's ecological environment is more and more concerned. The hosting of 2010' EXPO provides Shanghai the opportunity to improve the city's environment. The site locates to the north of Huangpu River and is closed to the city center. The aim to transform it from a closed waterfront industrial area to an open urban green space will accomplish through this design. Our scheme starts from the environment of wind and water, which are the most important elements of the local microclimate, and 'eco-layer', a tri-dimensional green space, is introduced. A series of open spaces is created to lead or gather wind by buildings and plants. Software is used to testify and optimize the scheme. Rainwater collecting system is applied to organize the water system in the site. Water volume achieves balanced. Water is activated both scenically and ecologically. The ecological functions of waters and aquatic plants are utilized to regulate the microclimate. Two water-belts link the historical context, which conserves not only the industrial buildings, but the industrial spirits of Jiangnan Shipyard and even the national industry in Shanghai. Apart from the aquatic plants for water purification, vegetation is covered in the site based on the cultural requirement, the arrangement of wind and water, as well as the treatment to the industrial pollution. The tense urban space is released through this reconstruction.

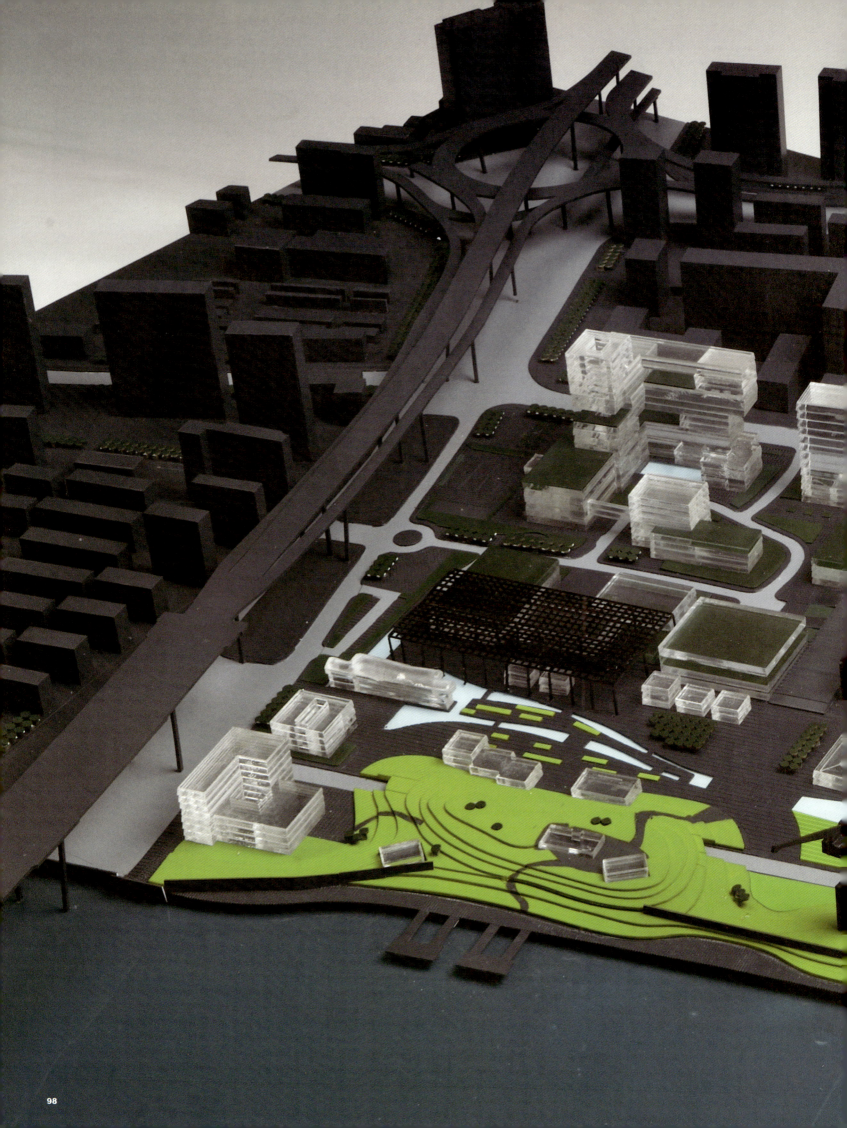

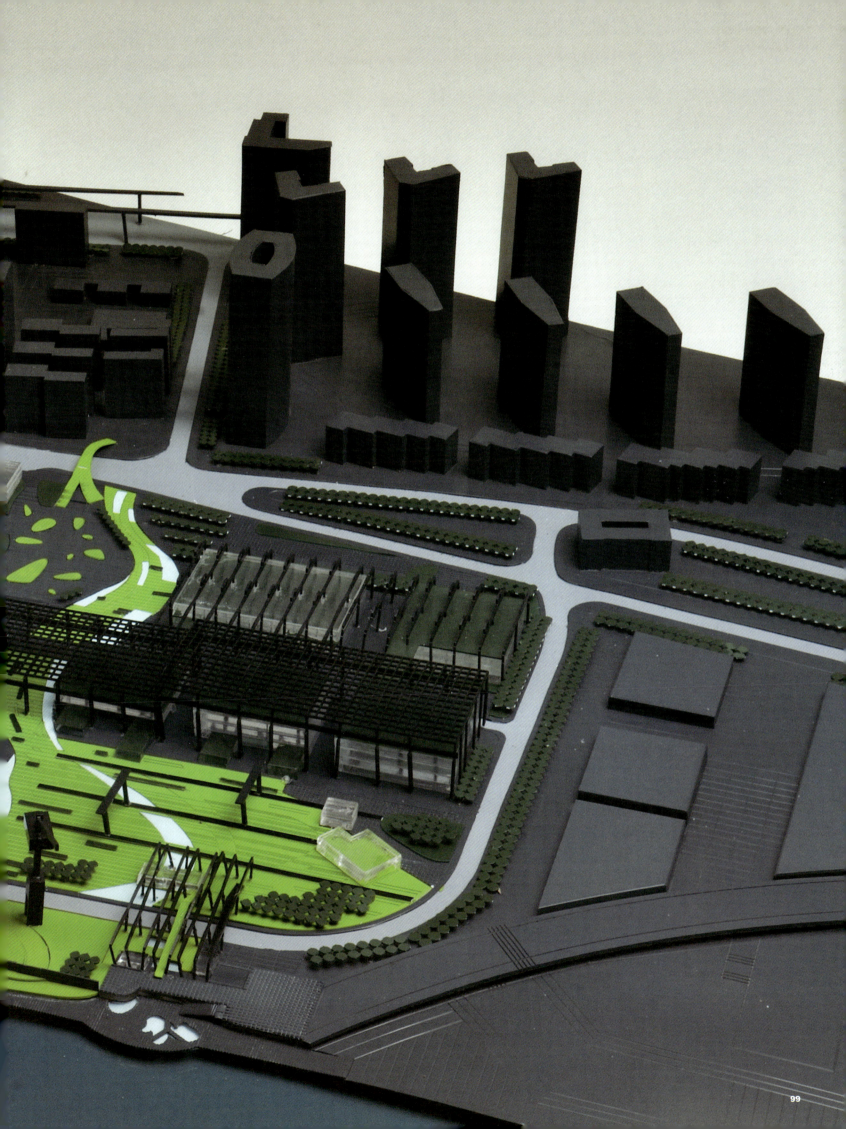

# 3 城市设计 Urban Design

### 业态分布
- 商业办公区（对外）
- 博览区
- 商业区
- 滨水区

### 高度控制
- 80m
- 30m
- 15m

### 车行交通
- 城市机动车道
- 区域机动车道
- 区域人车混行道
- 消防车道
- 地面停车
- 地下停车

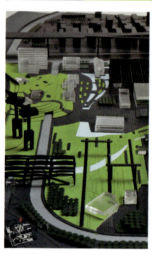

### 步行交通
- 城市交通下客点
- 沿街步行道
- 区域主要步行道
- 人车混行道
- 区域其他步行道
- 开放空间

## 旧建筑改造方式

**屋面改造：**
太阳能光电板的使用

**空间改造：**
大尺度厂房下建造小尺度新建筑，新旧建筑交错形成灰空间，新旧屋顶之间覆以生态层。

**表皮改造：**
增设通风井，利用自然热压改善飞机库内的生态环境。

**空间置换：**
开放厂房空间，利用原有结构营造外部公共空间。

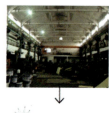
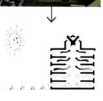

改造意向

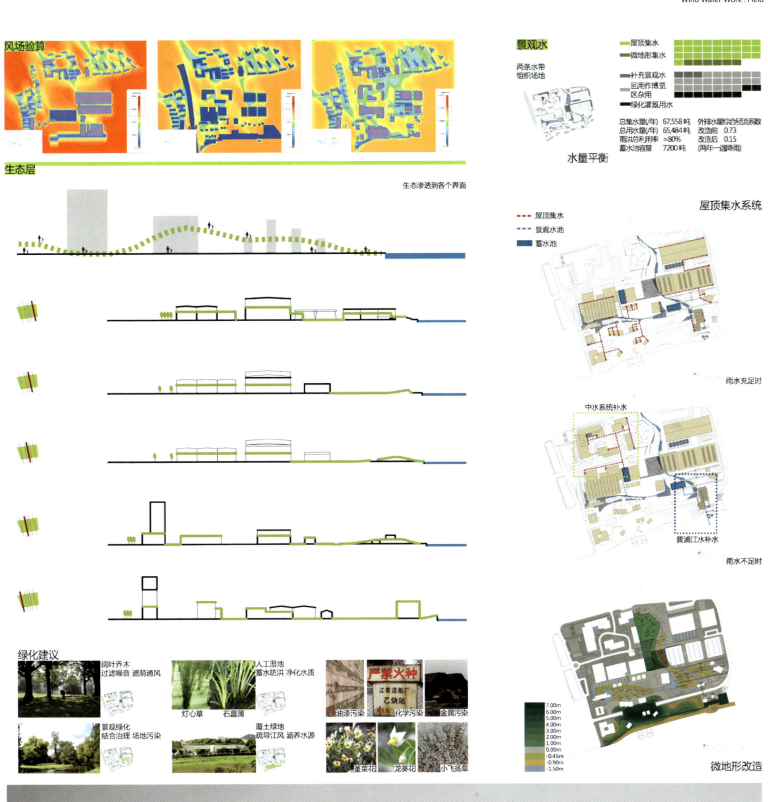

## 活水工厂

单体设计 > 活水工厂 > 设计：姚昕悦 江天 何晋

Architecture > Flowing Water Factory

管子工厂不仅联系江南造船厂和滨江，且是连接黄浦江和场地内景观水池的枢纽。方案从水出发，创造了一条水循环的路线，利用原有建筑的结构和残片，在景观和生态两方面活水，营造了一座活水工厂，同时整个净水系统的展示也成为场地改造后工业博览群中的主题场馆之一。

对原建筑的利用
屋顶

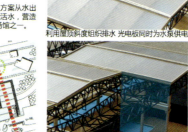

利用屋顶斜度组织排水 光电板同时为水泵供电

桁车

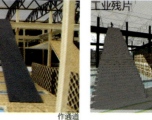

作通道

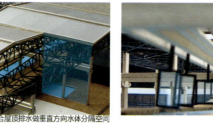

结合屋顶排水做垂直方向水体分隔空间

作景观

工业残片

作隔断

作小品

风场·水场·工场

# 汽车主题博物馆

单体设计 〉汽车主题博物馆 〉设计：袁懿 李邦健

Architecture > Autocar Museum

在中国，船的辉煌已成为过去，
这是一个属于汽车的时代.
我们何去何从

## 区位

## 概念

## 展示系统

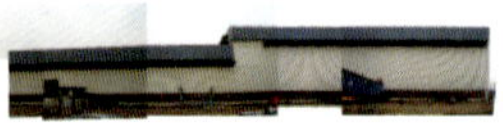

现有厂房

单向的厂房空间，往返的来回流线

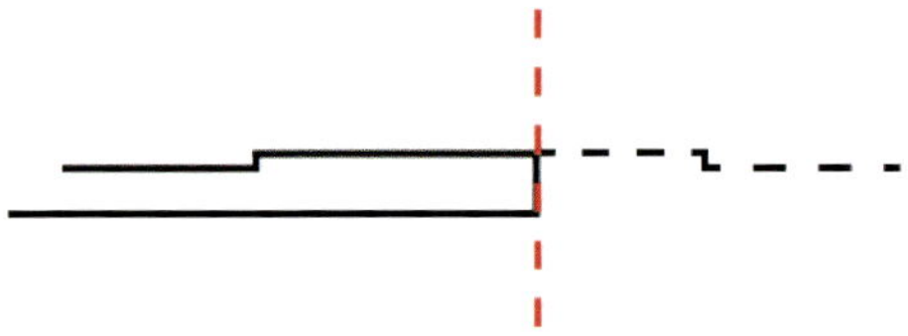

"MIRROR!"连续的展示体验，延续的厂房空间

贯穿公共交通流线

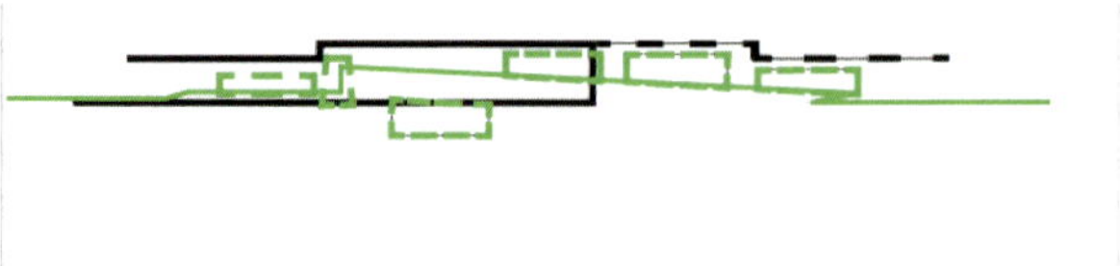

插入风，水，工厂特色体验空间

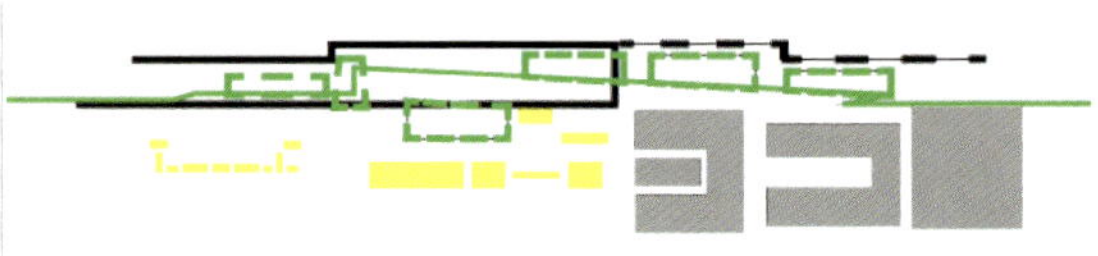

设置特色适宜的展示空间

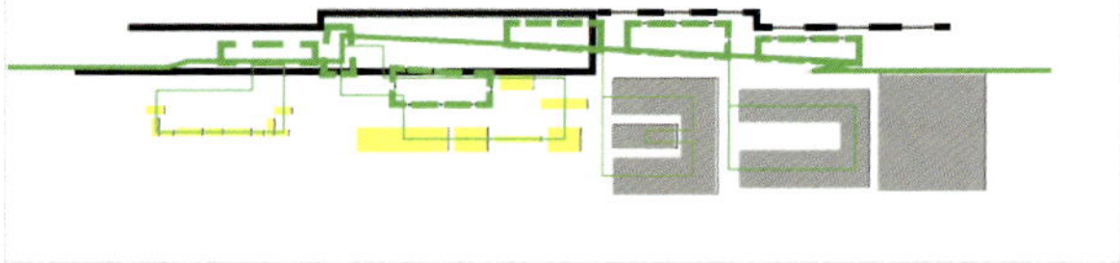

将展示空间链接入特色体验空间

## 体验系统

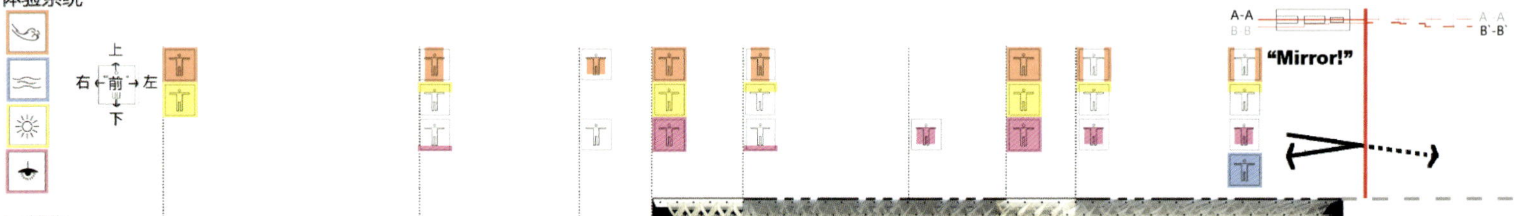

上
右 ← 前 → 左
下

A-A剖透视

"Mirror!"

TURNPOINT--
END
&
START

B'-B'剖透视

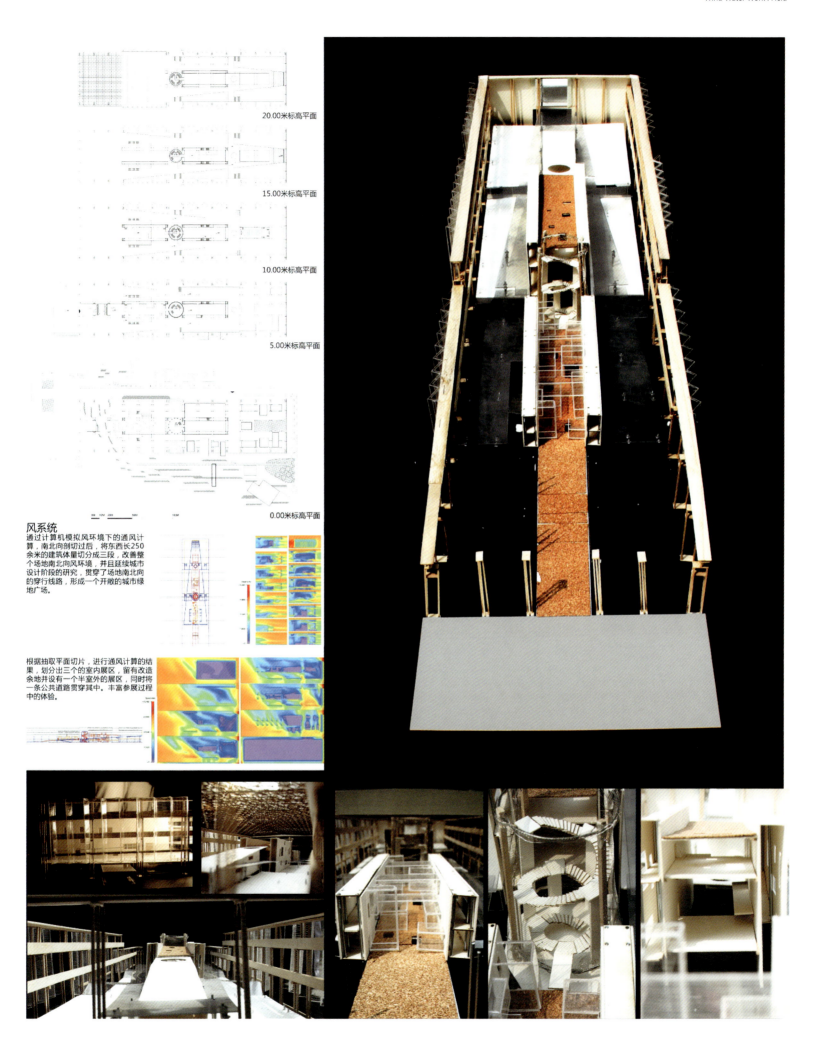

20.00米标高平面
15.00米标高平面
10.00米标高平面
5.00米标高平面
0.00米标高平面

**风系统**
通过计算机模拟风环境下的通风计算，南北向剖切过后，将东西长250余米的建筑体量切分成三段，改善整个场地南北向风环境，并且延续城市设计阶段的研究，贯穿了场地南北向的穿行线路，形成一个开敞的城市绿地广场。

根据抽取平面切片，进行通风计算的结果，划分出三个的室内展区，留有改造余地并设有一个半室外的展区，同时将一条公共道路贯穿其中。丰富参展过程中的体验。

张玉坤　　徐苏斌　　青木信夫　　陈天　　谭立峰

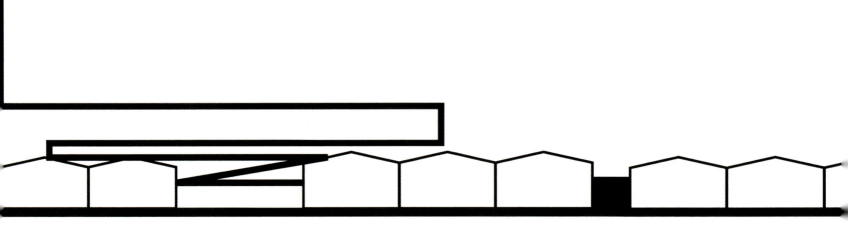

马睿　王特立　唐涵　李程　任桂园

天津大学

廖胤希　张晓未　惠超微　何美　孙洋

向迪士尼学习

## 世博背景概览

| 1970 | 1992 | 2000 | 2005 | 2010 |
|---|---|---|---|---|
| 日本大阪 | 西班牙塞维利亚 | 德国汉诺威 | 日本爱知 | 中国上海 |
| 64,318,770 | 41,814,571 | 18,000,000 | 22,049,544 | ? |
| + | - $ 0.3Billion | - $ 1Billion | 展后材料废置 | |
| 促进了日本关西经济带的形成 | | 奠定汉诺威全球会展龙头城市的地位 | | |

城市生活 —— 世博会 —— 城市生活
2007 2008 2009 **2010** 2011 2012 2015 2020

综合类世博会的参观人数呈下降趋势，而通过世博直接盈利也变得越加困难。然而世博带动的**城市建设、经济发展作用**远比会议本身具有深远价值。因此，有关世博的城市建设、经济计划都应立足于**长远的发展**。**2010**中国上海世界博览会，**城市，让生活更美好**，一届立足于展现城市生活变化的世博会。我们应该通过怎样的方式去实现这一主题？

以世博为契机，定位于会后，为城市，而非单为世博，设计中国近现代工业博物馆群

## 场地区位分析

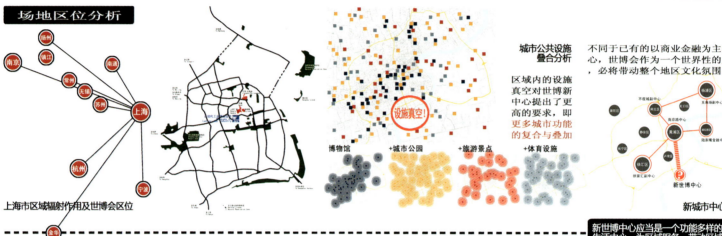

上海市区域辐射作用及世博会区位

城市公共设施叠合分析

区域内的设施真空对世博新中心提出了更高的要求，即**更多城市功能的复合与叠加**

博物馆　+城市公园　+旅游景点　+体育设施

不同于已有的以商业金融为主的城市中心，世博会作为一个世界性的**文化品牌**，必将带动整个地区文化氛围的提升。

新城市中心职能分析

新世博中心应当是一个功能多样的城市文化生活中心，为区域服务，带动区域发展起步

---

2007年国庆长假，武侯祠、成都工业文明博物馆参观人数为210,000 vs 4,900。
根据北京市免费开放的33家博物馆博物馆规模，假设每馆日人流量为2000人，全年开放时间为365x(6/7)=313天，北京市博物馆全年参观人数为 2000X313X33=2065万人

洛杉矶迪士尼乐园日均人流量为4万人，高峰时可达8万人，年人流量超过1800万人。
香港迪士尼乐园建成第一年人流量达到560万，预计未来五年内年人流量将超过1000万。
东京迪士尼乐园自建成后参观人数持续增长，两个主题公园的年人流量超过2500万。

事物的第一性是决定事物归属的决定因素，即**本质因素**；而第二性是其具有的特征，它不影响事物的本质，但是事物的重要**组成因素**。第二性是对第一性的**发展与补充**，也是展现第一性的重要**途径**。

博物馆的第一性：**展览教育性**
迪士尼乐园的第一性：**娱乐性**

在保证博物馆的展览教育性不变的前提下，研究促使迪士尼乐园兴盛的第二性要素，提炼调整后，应用的设计中，以促进博物馆第一性特征的最大发挥。

## 方案生成分析

### 建筑！ARCHITECTURE

通过分析，我们得出结论，为保留建筑的原真性与历史记忆，新加建筑面积不得超过保留建筑的建筑面积，即**新旧比不大于1**。

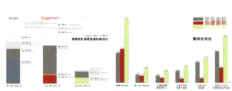

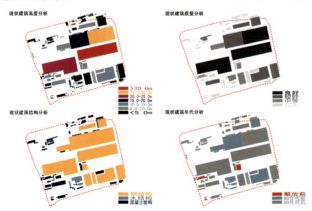

现状建筑高度分析　　现状建筑质量分析
现状建筑结构分析　　现状建筑年代分析

### 工业景观！INDUSTRIAL LANDSCAPE

中心工业遗存　分数工业遗存　部件装配车间　船体联合车间
中心工业遗存　分数工业遗存　国民党海军车间令部　西区装焊工场
滨江工业遗存　分数工业遗存　飞机库　管子工场

建筑保留方案分为**保护、保留**与**拆除**三个等级。保护具有重大历史价值的、建国前建造的建、构筑物。保留具有较强工业景观价值的、现状质量较好的、有改造再利用价值的建、构筑物。拆除质量较差的、景观价值不强的、影响重要建筑物视野的建、构筑物。

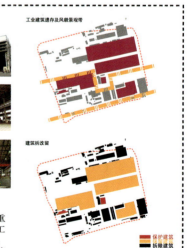

工业建筑遗存及风貌景观带

建筑拆改留

通过对**建筑结构、高度、建造年代、现状质量**以及**工业景观价值**的分析，确定加护保留建筑的原则。

 **天津大学** > 向迪士尼学习 > 设计：**马睿 唐涵 王特立 李程 任桂园**
Tianjin University > LEARN FROM DISNEY

# 向迪士尼学习 Learn From Disneyland

**上海江南造船厂 — 中国近现代工业博物馆群城市设计**　Urban Design of Chinese Latter-Day Industry Museums

　　一届世博会给城市带来的利益并非只局限于直接经济利益，它对城市发展的影响是巨大的，应该从长远发展的角度看待世博，而城市设计亦是如此。本设计以世博为契机，定位于会后，为城市，而非单为世博，设计中国近现代工业博物馆群。区位上的设施真空现象需要这里成为一个功能复合的生活中心，而世博会的召开将提升区域的文化氛围，这样，我们的博物馆群将同样担负着城市文化生活中心的职能。

　　在了解社会对博物馆的普遍态度后，我们反思博物馆现存的问题，内容枯燥、形式单一等等使大多数博物馆无人问津，而兴盛数十年之久的迪士尼乐园依旧吸引着大量人流。我们剖析隐藏于其本质属性娱乐性后的诸多因素，便捷的交通，充满个性的建筑与景观，完备的配套设施，多样的主题功能，充满参与性与体验性等，是使其经久不衰的深层次原因。我们将向迪士尼学习，在对其第二性要素研究提炼后，应用到我们的设计中去，在设计中充分利用周边便捷的交通，合理组织内部交通；处理与利用好工业建筑与遗存，使其成为基地特色之一；研究功能混合形式，合理地将城市文化生活中心与博物馆群的功能复合化，达到高效利用；创造充满体验性与参与性的空间，达到吸引人流的目的，创造一个新型的、充满活力与吸引力的复合型博物馆文化生活中心。

　　What will be bring to the city by holding an EXPO？When everyone pay attention to the money, the influence to the city development is expanding. So we should value the EXPO with a long view, so is the urban design. This time, we design the museums for the city, not only for the EXPO. There is a big lack of public facilities in this area, a public plaza with mixed functions is wanted. And the EXPO will bring much cultural atmosphere here. So it's a god-given chance to the site.

　　People are boring with museums for its unchanging, homiletic, etc. So a big mount of museums are out of use. However, the fifty-year-long Disneyland is still be welcome. Why？ First of all, it's a amusement park, people go for amusement. It's different of museums. But what elements made it so successful？It must be something deeper. By researching all-sidedly, we found the reasons：a perfect transportation, uncommon buildings and sights, consummate serving and all-sided functions, different themes, joinable games and wonderful experences. We will learn from Disneyland, using the elements into our design. We will make full use of the efficient transportation of Shanghai, and connect the interposition rationally. And we should make full use of the huge industrial buildings and lefts, make our site different to others. In our design, we divided the function elements of museums and plaza, and mixed them with rational connection. So it is designed by layers, and become much efficient. We made pipes in our design. In the pipes, you can get the information in all different ways, and it can make you much excited. Enough open spaces are made for the possible games.

# 交通！TRANSPORTATION

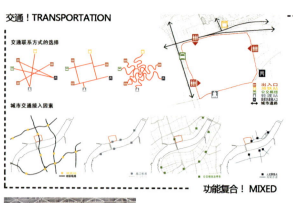

## 功能复合！MIXED

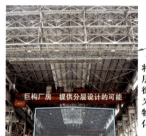

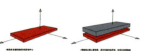

将通达性强、地块价值高的地平层设计为**市民文化生活中心**，沿街优先布置商业建筑，并将其作为现状和限制条件构成立体的博物馆基地，形成开放式的、生活化的博物馆

市民文化生活中心为博物馆群**提供服务支撑**，博物馆群为文化生活中心**提供文化支撑**，二者相互为对方吸引人流，成为良性循环系统。同时，通过政策调节，用商业性功能区的部分**租金支持公益性**的博物博览业的发展，可减轻社会的经济负担。

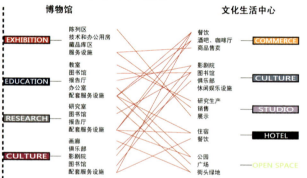

方案生成分析　MARUI DESIGN

## 主题分区！THEME

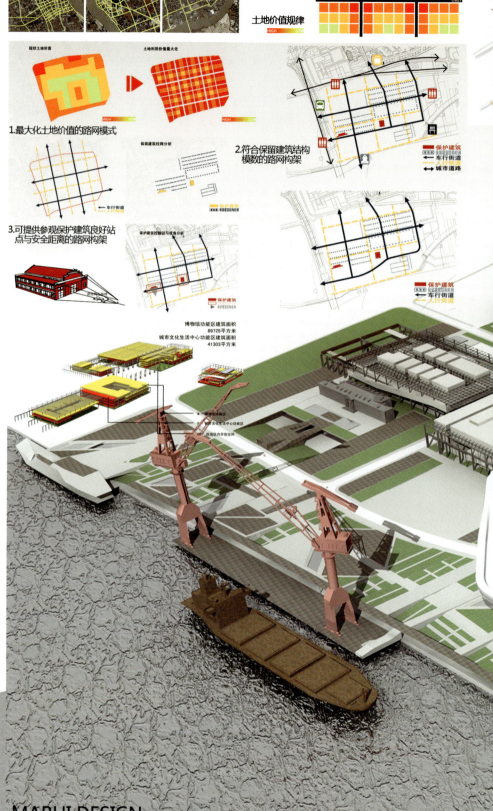

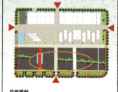

## 体验！EXPERIENCE

## 参与！JOIN

参与 - 场地 - 开放空间生成

我们将参与性的体现划分为建筑内部与场地中两部分，通过提供活动发生空间的方式创造参与的可能性，并实现建筑内外的互动。

### 管的生成

1 中心工业遗存轴与滨江工业遗存轴
2 博物馆主入口轴达到博物馆层主入口，城市生活中心主入口轴深入基地内部
3 入口广场，保护建筑广场形成三个主要开放空间
4 滨江工业遗存轴扩展为滨江休闲区，联系两座保护建筑广场，入口轴、遗存轴将广场串联起来
5 各地块内小型开放空间
6 主体结构与小型开放空间形成开放空间结构系统

### 经济技术指标

改造区博物馆建筑面积 89725m²
改造区文化生活中心建筑面积 41303m²
管空间博物馆建筑面积 20760m²
新增文化生活中心建筑面积 78731m²
博物馆总建筑面积 110485m²
文化活动中心总建筑面积 120034m²
保留与改造总建筑面积 131028m²

新增总建筑面积 99491m²
新旧比 0.76
基地占地面积 234356m²
总建筑面积 230519m²
建筑密度 36%
容积率 0.98
绿地率：28%
停车位数量 520个

通过动态流组织的博物馆群参观主流线

城市文化生活中心层组织的博物馆交通流线

### 城市设计导则

**建筑改造数据：**
主体结构占地面积 10543m²
建筑面积 22951m²
博物馆部分面积 16186m²
生活中心面积 6765m²
M/P-t  5-5.5
M/P-b  0.2-0.4
OS  0.2-0.25

**地块数据：**
占地面积 21965m²
容积率 1.04
建筑密度 48%
建筑限高 25m
绿化率 30%

**地块设计要求：**
建筑设计与景观设计紧密围绕轻工业这一主题。处理好管子与建筑的衔接。注重建筑内与场地上参与空间的创造。

**建筑改造数据：**
主体结构占地面积 6848m²
建筑面积 13410m²
博物馆部分面积 8680m²
生活中心面积 4730m²
M/P-t  4-4.5
M/P-b  0.3-0.4
OS  0.4-0.5

**地块数据：**
占地面积 23235m²
容积率 0.6
建筑密度 30%
建筑限高 25m
绿化率 35%

**地块设计要求：**
建筑设计与景观设计紧密围绕世博历史这一主题。处理好改造建筑与保留历史建筑的衔接。注重建筑内与场地上参与空间的创造。

## URBAN GRAFT

### Diagram

**Modulate**
the pipe in accordance to the factory's structure

**Connect**
two parts of the pipe by extensions

**Modify**
the slope angle to 30 degrees so as to make escalators

**Reinforcment**
of the pipe by adding stick column to the bottom

### Sun Analysis

Shanghai's sun altitude:
83 degrees (summer)
36 degrees (winter)

A comfort light environment is indispensable for displaying art. The side window always leads to direct glare.

The traditional museum use skylight to optimize the incident light for a better appreciation of arts, but it could also cause reflected glare.

Then we use both skylight and curved walls to manipulate the incident light, most of which would be second reflected light that would be diffused and soft for visualization.

### Exhibition Analysis

Best vertical eye angle 26 degrees

### Architectural Concept and Urban Strategy:

The new museum for industry addresses the question of its urban context by maintaining the former shipdock. This is in no way an attempt at topological pastiche, but instead continues the large pattern shipdock texture set against the small pattern urban texture on the surrounding sides of the site. In this way, the museum is more like an 'urban graft', a second skin to the site.

The entire building has an urban character: prefiguring upon a directional route connecting the site to Huangpu River, the museum encompasses both movement patterns extant and desired, contained within and outside. This vector defines the primary entry route into the building. By interwining the circulation with the urban context, the building shares a public dimension with the city, overlapping tendril-like paths and open space. In addition to the circulatory relationship, the architectural elements are also geometrically aligned with the urban grids that join at the site.

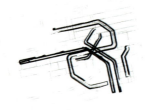

# URBAN GRAFT

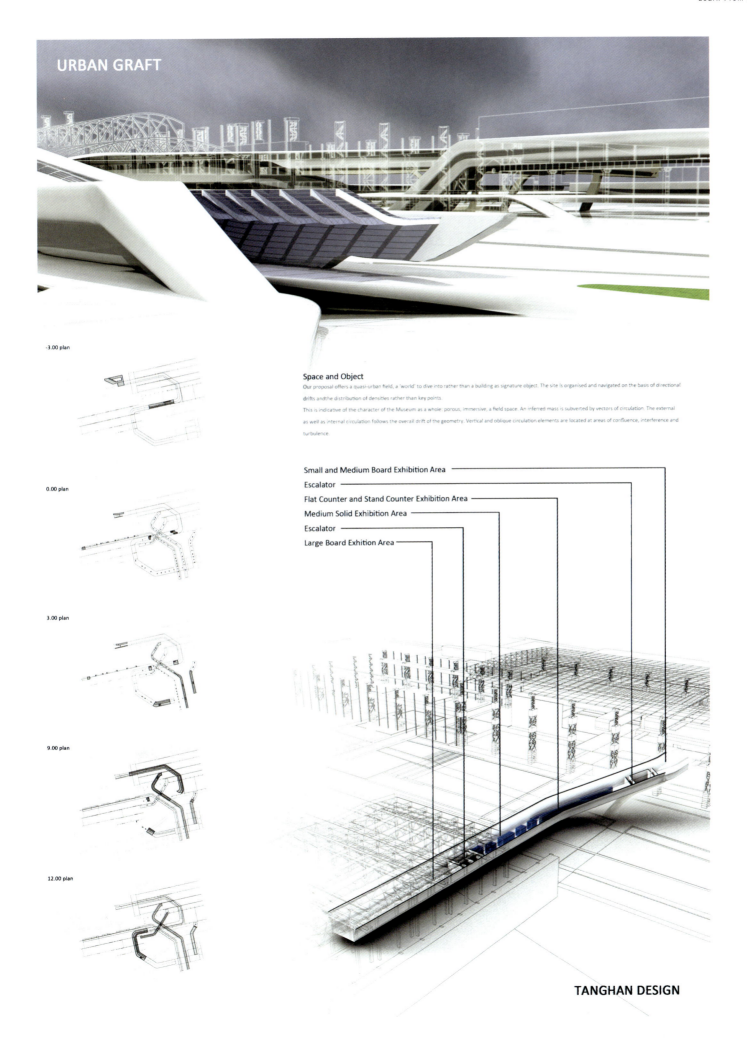

-3.00 plan

0.00 plan

3.00 plan

9.00 plan

12.00 plan

### Space and Object

Our proposal offers a quasi-urban field, a 'world' to dive into rather than a building as signature object. The site is organised and navigated on the basis of directional drifts and the distribution of densities rather than key points.

This is indicative of the character of the Museum as a whole: porous, immersive, a field space. An inferred mass is subverted by vectors of circulation. The external as well as internal circulation follows the overall drift of the geometry. Vertical and oblique circulation elements are located at areas of confluence, interference and turbulence.

- Small and Medium Board Exhibition Area
- Escalator
- Flat Counter and Stand Counter Exhibition Area
- Medium Solid Exhibition Area
- Escalator
- Large Board Exhition Area

**TANGHAN DESIGN**

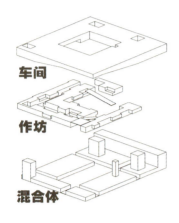

车间
作坊
混合体

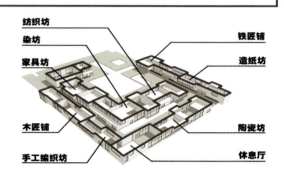

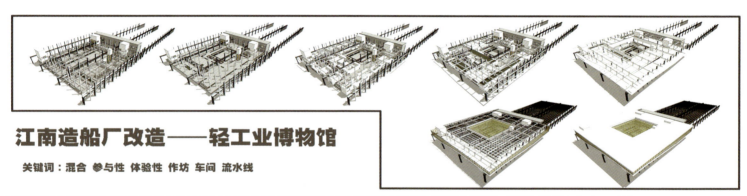

# 江南造船厂改造——轻工业博物馆

关键词：混合 参与性 体验性 作坊 车间 流水线

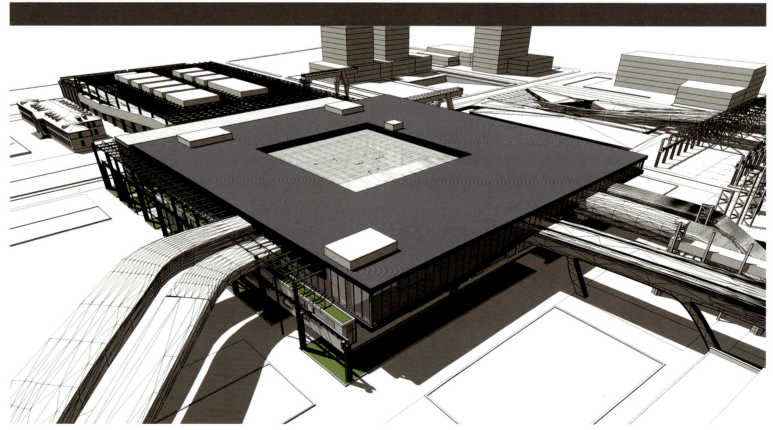

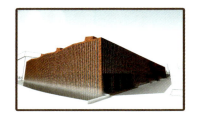   

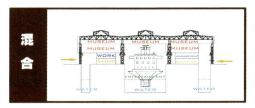 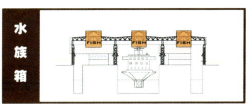 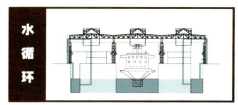

混合　　水族箱　　水循环

空中构筑物

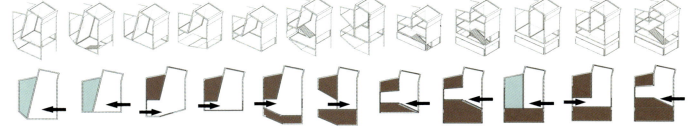

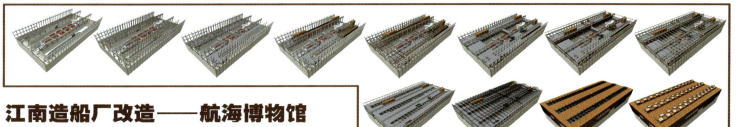

## 江南造船厂改造——航海博物馆

关键词：混合 体验性 娱乐性 船舶 海洋 船舶工业 水

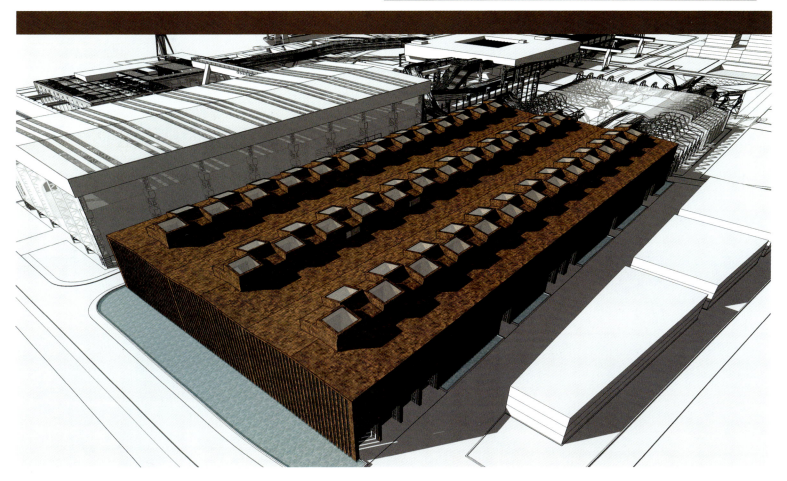

**沿江景观总平面图**
LANDSCAPE MASTER PLAN OF ALONG THE RIVER

规划理念：

景观的实用性：景观空间首先要满足需求与功能，同时与基地环境相互协调，能容纳公众多种活动，从物质上和精神上去引导居民的正常生活

景观的多样性：在保护环境、健康发展的前提下，提供多样化的环境空间，让优美的环境景观与周边区与环境共生共荣，和谐统一

景观的延续性：在建设中保持与周边自然环境，城市文明的延续性

景观的艺术性：运用各种建筑手法和当地自然的材料、植物，用艺术的表达构成人文自然景观，使生活、活动在其中的人获得艺术享受

硬质铺装（Hard--Scape）
主要道路（Main--Street）
景观水系（Landscape Water）
硬地广场（Hard Floor Square）
绿地（Greenery）
步行路线（Pedestrain Path）
景观桥（Landscape Bridge）

**景观分区总平面图**
LANDSCAPE SUBAREA PLAN

# 破土：超越时间的新生

图解基地城市事件

需求与空间的矛盾

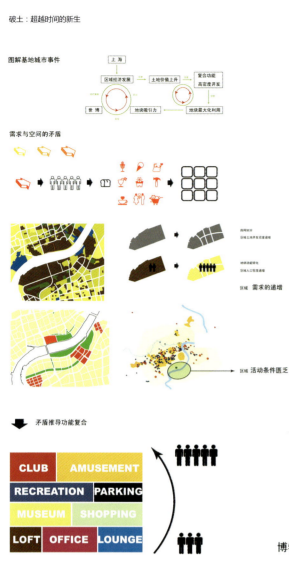

矛盾推导功能复合

博物馆自身文化挖掘

文化挖掘——解读上海城市缺失

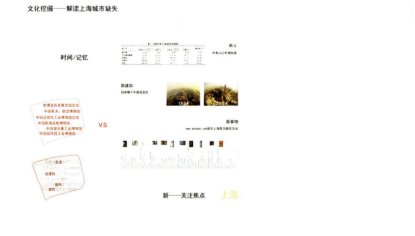

博物馆改变物质排序方式

基地功能分布示意图

方案定位演化示意：

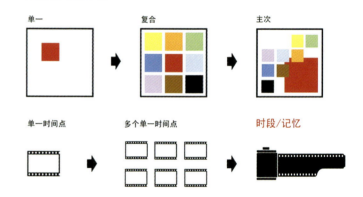

基地功能演进示意图

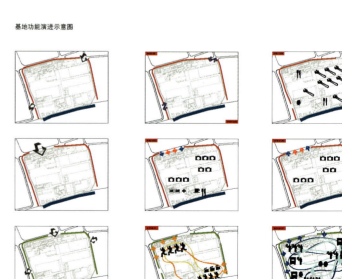

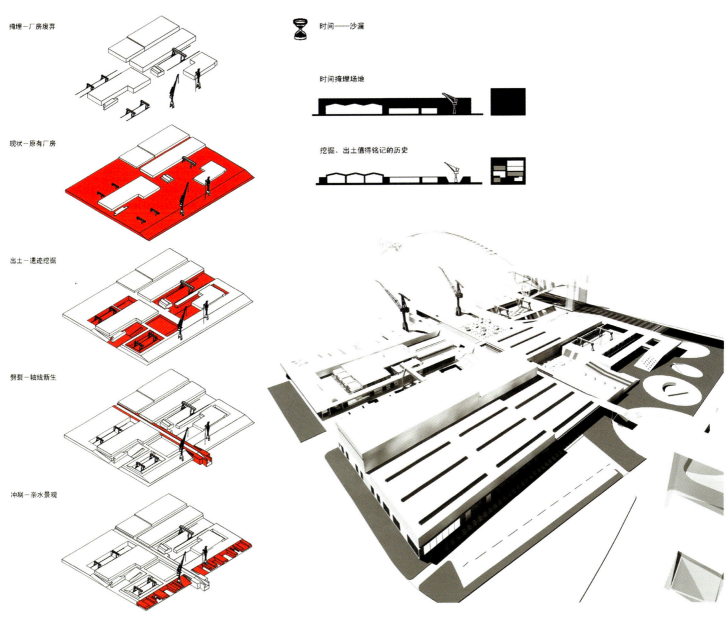

天津大学 > 2010年上海世博会中国现代工业博物馆群设计 > 设计：廖胤希　张晓未　惠超微　何美　孙洋

# 破土：超越时间的新生
## Breaking Ground: The Rebirth Transcended Time

　　本设计从大都市城市记忆缺失这一现象出发，通过对城市工业的片段整理，重新梳理时间序列，以一种反思的态度将近代工业的历史、旧有厂房在工业污染之后废弃掩埋、破土而出的全部过程，跨越时间与空间的限制，展现在城市之中，从而超越历史现象，回归事件本质，使人们重新审视现有的生存空间。

　　我们将旧有厂房以近代工业博物馆为功能，作为城市工业记忆体"掩埋"于新市民广场之下。通过"挖掘"，使部分厂房与设备破土而出，使得人们在新市民能够感受到宛如考古现场的超现实主义景象，达到了历史与未来的对话。

　　博物馆与新市民广场共同构成了以时间为线索的游历序列，展示了整个近代工业的兴衰史和发人深省的未来景象。设计通过垂直穿过基地的中轴线，连接作为主要展览空间的旧有厂房。我们将参观过程中历史事件的时间序列与参与者的心理序列作为设计的出发点，使参观者经历工业萌芽期的欣喜，工业秩序期的规律，工业膨胀期的压抑，工业衰落期的失落。在轴线的终端以黄浦江水域断面作为参观序列的高潮，使人置身水下，以一种灾难的场景令人自省，穿越时空与自然对话。最终，人们回归新市民广场，审视现状，思考未来。

　　This design from the phenomenon of memory loss in the metropolitan cities, through the re-organization of industrial region, re-organizes time series to be a reflection of the attitude of modern industrial history. the old plant in the industrial pollution abandoned after the burial, ground-breaking and To the entire process, across time and space constraints, demonstrated in the cities, so as to transcend the historical phenomenon, the return of the incident nature so that people can re-examine the existing living space.

　　We will have the old plant to the museum for modern industrial function, as the industrial cities of memory "buried" under the new public square. Through the "mining", so that part of plant and equipment Potuerchu, making people in the new members of the public can feel like a archaeological site of the surrealistic scene, to the history and future of the dialogue.

　　Museum and the new public plaza constituted a common time for clues to the travel sequence, showing the entire history of the rise and fall of modern industry and thought-provoking picture of the future. Through the design of the base through the vertical axis linking the main exhibition space as the old plant. We will visit the historical events in the course of the time series with the participants of the sequence as a psychological point of departure designed to make visitors experience the joy of budding industry, industrial order of the law, industrial expansion phase of depression, the loss of industrial decline. Axis in the terminal waters of the Huangpu River to visit the section as the climax sequence, people outside the underwater to a disaster scene is self-examination, through time and space and natural dialogue. Eventually, people return to the new public square, look at the status quo, think about the future.

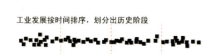
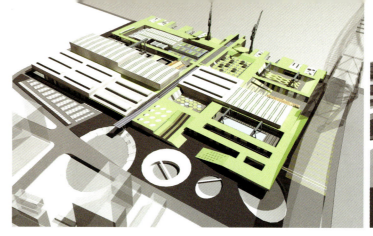

工业发展按时间排序,划分出历史阶段

不同历史阶段进驻不同展区

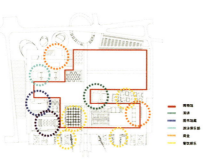
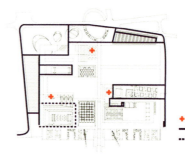
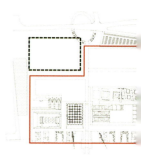

方案城市设计分析

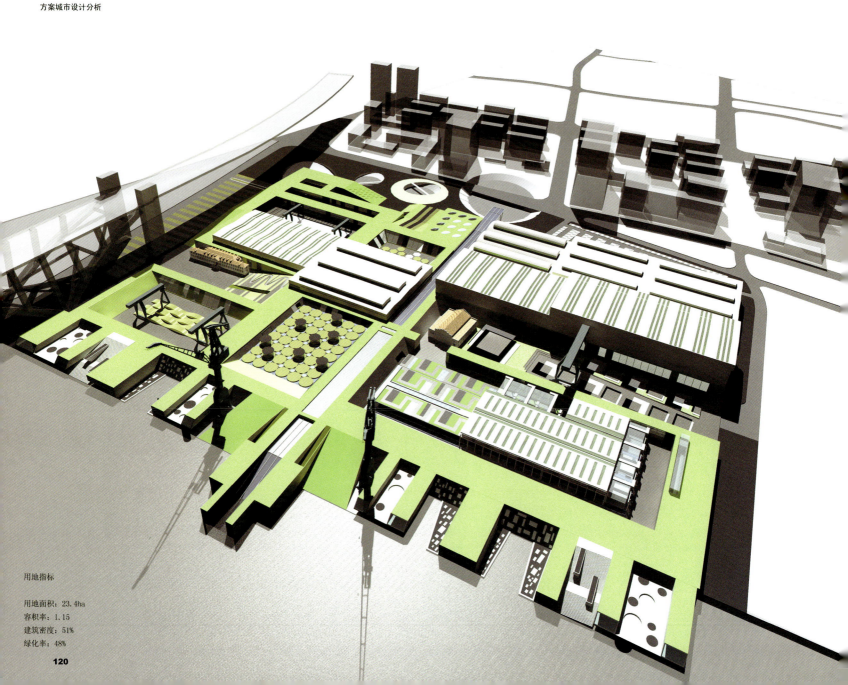

用地指标

用地面积: 23.4ha
容积率: 1.15
建筑密度: 51%
绿化率: 48%

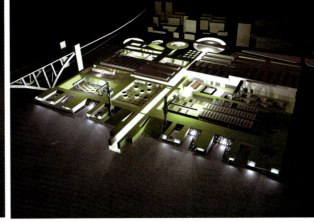

### 规划设计
方案关注城市区域发展需求，充分利用城市规划相关交通、基础设施条件，进行复合功能开发和市民活动空间的提供。基地公交系统便捷，内部交通环境充分考虑步行系统的舒适性，最大化的提供多样的市民活动空间。城市设计划分了分期建设区域，对基地经历世博会作出了相应的设计。

### 建筑材质比的研究
通过对基地旧有非保留建筑的材质配比研究，我们发现有很多部分的废弃建筑材料可以回收利用，我们将这些建筑材料分为预制板、抹面、砖、钢材、木等类型，通过计算得出用料比为预制板:抹面砖:钢材:木=3:4:3:1，立面材质比为抹面:砖:钢材:木=5:1:3:1。我们通过重新设计，将建筑材料重新整合，重新创造出多种建筑立面模块，用于对旧有建筑的外立面"掩埋"。

### 珊瑚馆设计
珊瑚馆位置接近中轴线终端，通过对其结构设计，使得其上部作为有屋顶种植的休憩空间及茶室，其下部可以看到上部所种植的树的根部，使人联想到珊瑚，通过控制珊瑚馆室内的亮度及光色，使的人们好似置身于海底之中，通过海洋生态污染题材的展品的布置，使参观者对工业污染现状反思。

### 滨河设计
通过对潮滩形态的研究，我们将其形成过程进行图解，并将最终结果应用于滨河景观设计当中，深入水中的建筑体量与广场渡口相结合，满足不同功能的需要，并且构成了如同潮滩一般的市民亲水景观带。

滨河设计图解

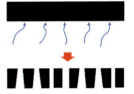

基地建筑材质比分析

珊瑚馆设计图解

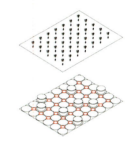

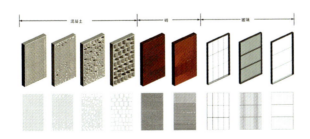

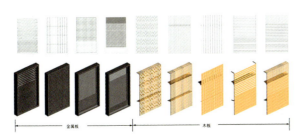

立面单元生成

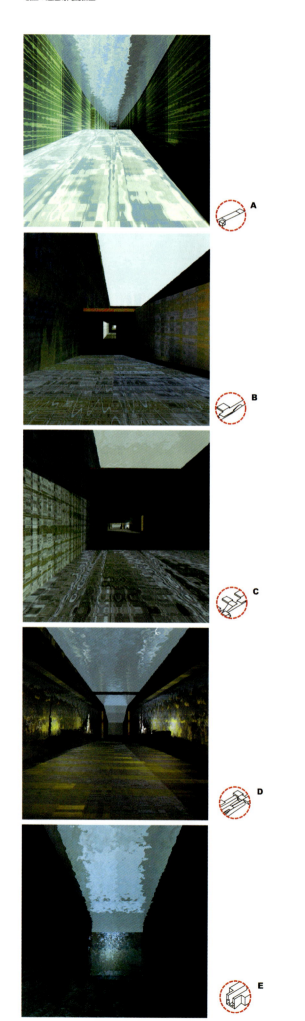
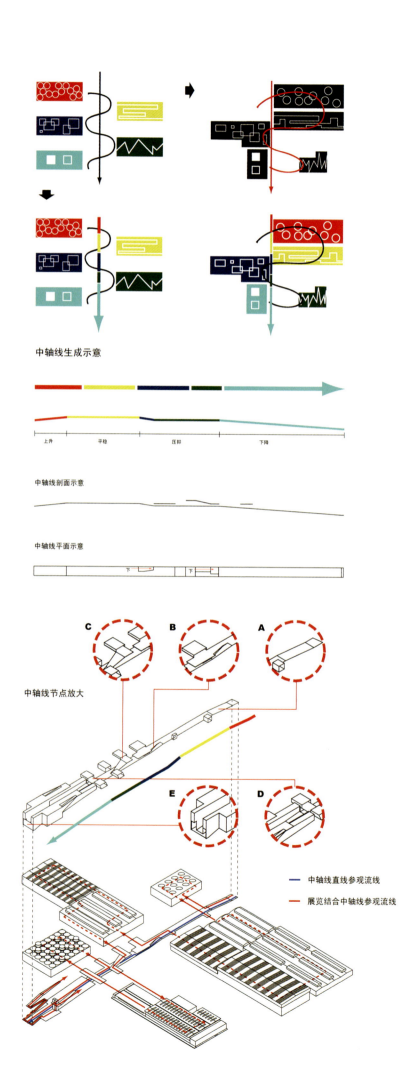

Breaking Ground: The Rebirth Transcended Time

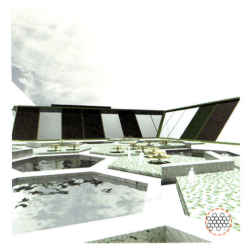 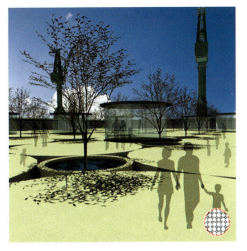 

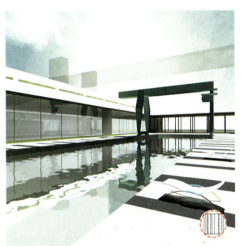  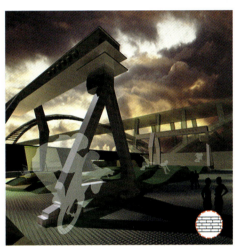

肌理化院落：

旧有厂房破土而出，与新市民广场形成院落空间，我们希望通过对活动与行为的归类，赋予院落不同功能，从而推导院落的空间形式。通过研究，以极限运动、水上娱乐、咖啡休憩、茶室餐饮、室外展示、表演聚会六种不同的活动形式为主题，推导出六种不同肌理的院落空间。使得每一个院落以一种统一简单的建筑语汇，能够容纳其功能并赋有个性。

活动产生形式

行为　＋　场所　➡　场所形式　➡　平面肌理

院落肌理分布图

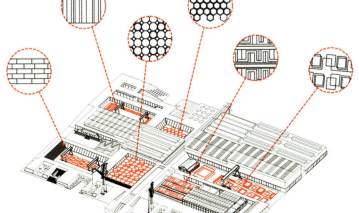

123

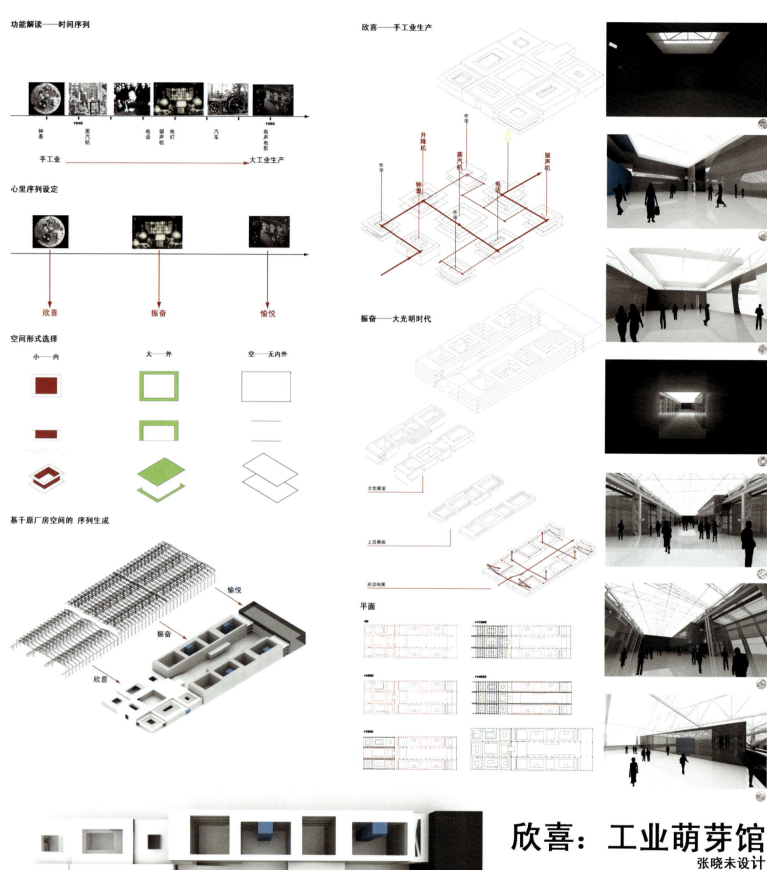

# 欣喜：工业萌芽馆
### 张晓未 设计

设计从时间序列出发，引导心理序列的产生，展示从手工业到大工业时代的历史阶段：纷繁的小发明指引欣喜，电灯的出现开启光明时代引人振奋，有声电影改变人类生活产生愉悦。基于该心理序列，选择以尺度的大小、内外、开合的不同体现空间属性，以明暗反差、封闭空间的转换划分不同心理区间，以虚的手法，留白展示原有厂房的体量。

## 秩序：工业时代馆

张晓未设计

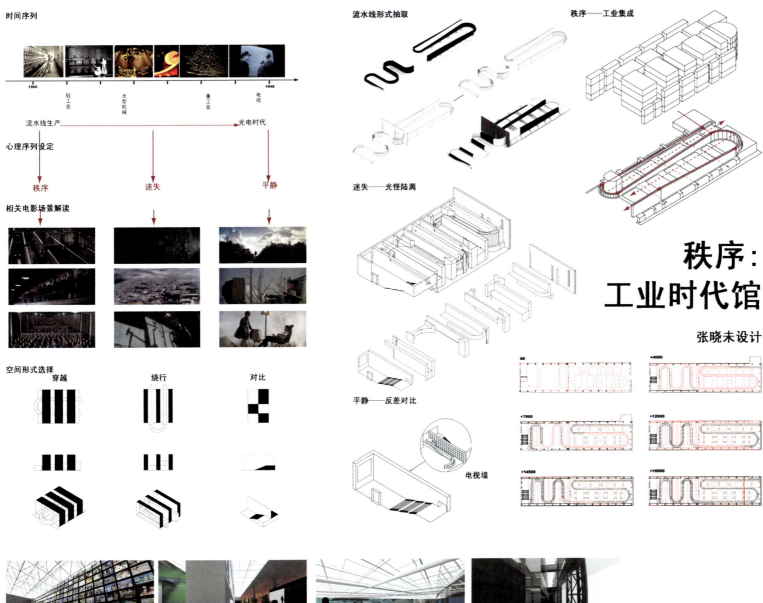

由大工业生产进入光电时代，流水线体现秩序化，大工业膨胀的纷繁复杂产生迷失，最终收结于电视回归平静。提取传送带作为元素，以穿越重复体量体现秩序，绕行纷繁界面体现迷失，嘈杂与空旷的强烈对比指引平静。最终用实填补了原有厂房的空洞。

破土：超越时间的新生

## 社区图书馆
### ——将优秀历史建筑还于市民

优秀历史建筑是属于市民的财富，而海军司令部与部件装焊车间的夹缝空间并不适合观赏整体建筑，将优秀建筑换一个角度去理解，它既可作为展品观赏，也可以作为借景形成一种历史文化氛围。

 **展品** 两建筑间形成展廊，透过玻璃幕墙，海军司令部背立面转化为展品观其细部

 **借景** 整体建筑的适合观赏点在社区图书馆内，此时海军司令部背立面作为借景延伸进图书馆内部空间

各层平面示意图

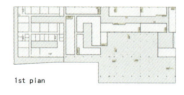
1st plan

2nd plan

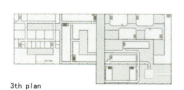
3th plan

## 博物馆展示区——工业膨胀下的心理序列

工业力量的逐渐膨胀使人愈显无限渺小。巨大非人空间尺度、高速运转的机器，机器vs人，我们的空间在哪里？

惠超微设计

**序列1**
工业发展阶段　巨型化
vs
人类心理感觉　自卑

失衡的体量
仰视的视角
尺度的错觉

巨型机械内部般的空间

**序列2**
工业发展阶段　密集化
vs
人类心理感觉　压抑

开阔
均衡
压抑

不同空间尺度下的心理感觉

**序列3**
工业发展阶段　高速化
vs
人类心理感觉　迷失

运动的影像展示——$V_1$
自动人行道——$V_2$
相对速度 $V=V_1+V_2$

单向传送管道作为封闭的交通暨速度体验空间联系各独立展厅

**序列4**
工业发展阶段　信息化
vs
人类心理感觉　空虚

材质体验——玻璃的处理
中空隔音——听觉——静
曲面磨砂——视觉——虚

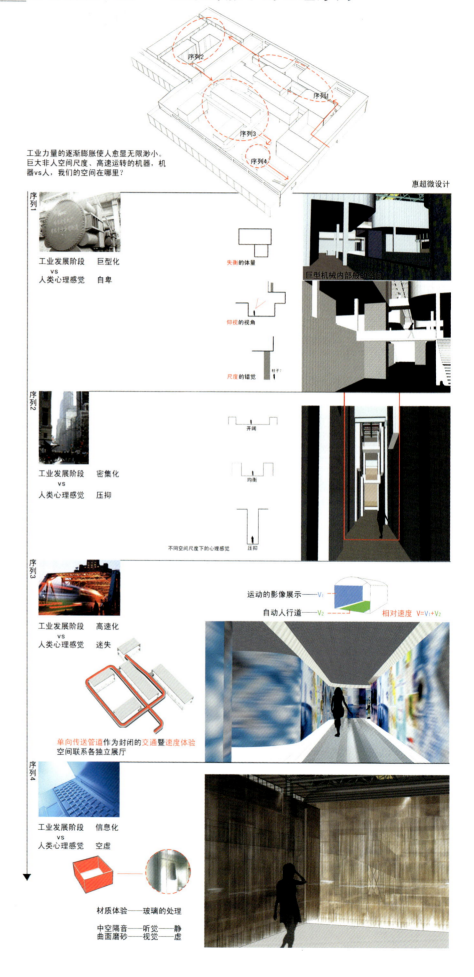

# 首层平面图

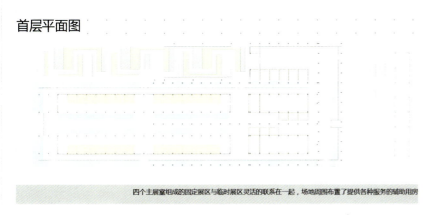

四个主展室组成的固定展区与临时展区灵活的联系在一起，场地周围布置了提供各种服务的辅助用房

## 海底时光隧道
—— 工业生态博物馆设计　　何美设计

游览者的视线穿过天窗表面摇曳的水波到达暴露出的结构和天车，廊道两侧的大型水族箱进一步营造出水下的氛围

从广场入口看飞机库

海下的亚特兰提斯神庙

因海平面上升而受到威胁的海岸

海上钻井平台

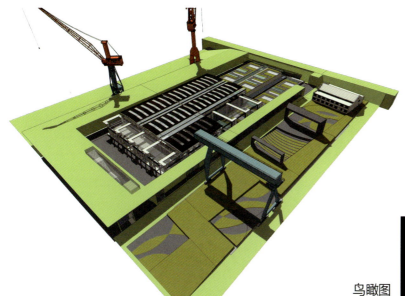

鸟瞰图

现有要素为待改造的工厂，围合感强烈的院子及贯穿基地的主轴；需要解决的问题为利用已有的结构和空间，以水的主题满足心理序列的整体需要并实现使用功能

现状

加入加建部分后工厂与虚实两种方式和主轴发生联系，根据原有结构将大块体量及加建部分进行分割并在由此产生的缝隙和建筑周边加入水系，代表"新"的玻璃体在入口处形成过渡空间进而侵蚀代表"旧"的厂房，新与旧紧密的咬合在一起。水的界面围合成廊道贯穿连接四个展室形成通畅的流线

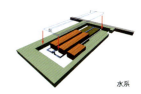

水系

规划后得到狭长的广场，历史保护建筑飞机库由于其自身重要性成为广场设计的起点，根据其所在的位置和朝向找到将基地划分为两部分的轴线，在轴线的一侧与飞机库产生对位关系的地方设置露天舞台，丰富多彩的交往活动将使这里成为广场上另一个不同性质的中心。在舞台两端的框状装置将使场地上人们的视线集中到飞机库的侧立面上，带状绿地景观区在轴线两侧波动，强调了轴线的存在，略有起伏的地形丰富了人们的心理体验；带状绿地的波动围合成硬质铺装的小广场，设计后的广场通过与周边场地发生交通和视线上的联系融入到周边环境中

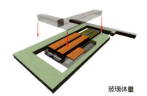

玻璃体量

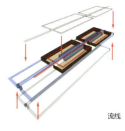

流线

在这个节点中，我们设想在遥远的将来，由于未能有效控制温室气体的排放，今天人们所担心事情变成现实，全球变暖使海平面上升，岛屿和大陆边缘的低海拔地区被淹没于一片汪洋之中并经历桑田沧海的变化曾经的繁华和混乱都将被静默无声的大海掩埋，如同古国亚特兰斯，成为又一处人类文明的遗迹，供后人吟咏嗟叹，感怀无限。从昔日机器轰鸣的喧嚣到眼前大海深处的宁静，我们相信这种强烈的反差一定会使游览者对工业与人，现在与未来产生进一步的思考，基于以上的思考和设想，我们开始了对这一部分的设计

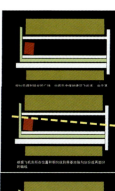

从入口处看过渡空间，以及厂房被改造后暴露出的结构和水的关系

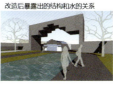

强调飞机库立面特征的门雕成为院落与广场的主要联结点，它是场地的标志也是中国近代工业的纪念碑

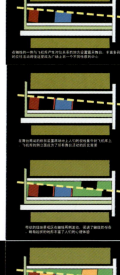

chongjian.com

FACULTY OF ARCHITECTURE AND URBAN PLANNING CHONGQING UNIVERSITY

王文婷　伍鹏晗　付烨　张琴　万博

王怡斐　韩文晶　崔璨　孙海霆　周雪峰

绿色城市

提出问题 Questions

1. 与城市功能脱离的封闭型空间

1. Enclosed space which is separated from the urban function.

2. 公共空间系统基本功能缺失

2. Lack of basic function of Public space systems

3. 现状防洪体系部分完全功能化，将人与浦江割裂

3. The present situation of flood prevention system is functionalizes completely, separating the people and Pujiang.

解决问题 Answers

1. 开放的空间，融入城市生活，为周边民众提供服务

1. The opening space, integrates the city life, provides the service for the peripheral populace.

2. 构建成体系的功能完善的公共空间系统

2. Construct perfect function of public space systems.

3. 合理利用防洪体系，改造为滨江湿地公园，实现人与亲水空间的相融合

3. Use the flood prevention system, transforming it to the wetland park, realizing the harmony situation of the people and the wetland park

设计构思 Design ideas

城市公共空间体系 system of urban public space  ------> 城市绿网体系 Green city network system

垂直于水体的街道空间是传统的城市自下而上的空间模式（因为通风和景观），这为实现向城市纵深方向渗透的绿色廊道的联网提供可能性，引导城市公共空间。

the vertical in the water body street space is the traditional urban space pattern, guiding the urban public space.

上海周边绿地演变图 Shanghai peripheral green space evolution chart

1980　　1986　　1998

上海世博会景观分析 landscape analysis of Shanghai World Expo

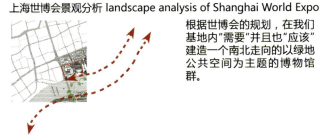

根据世博会的规划，在我们基地内"需要"并且也"应该"建造一个南北走向的以绿地公共空间为主题的博物馆群。

基地周边现状绿化
Peripheral afforestation

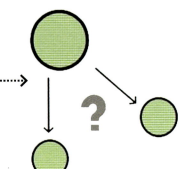

现有状态下的"孤岛"
isolated island

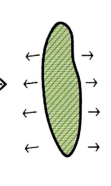

长条形扩大了绿色服务面
The long linearity expanded the green service surface

围和有特色的场所空间
encircle the characteristic space

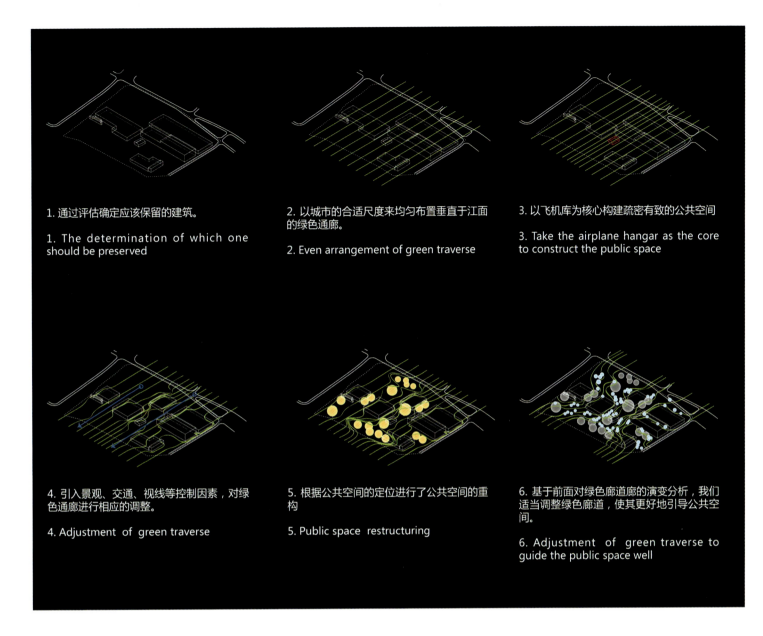

重庆大学 〉城市建筑的更新与再生--2010年上海世博会中国现代工业博物馆群设计 〉设计：万博　伍鹏晗　张琴　付烨　王文婷
Chongqing University >Regeneration of Urban Architecture-Museum Complex of Chinese Modern Industry for EXPO2010 Shanghai

# 绿色城市 Green City, Green Net

绿色城市 — 更美的城市 — 更好的生活 Green City-Better City-Better life

城市的发展，交通量的急速增长，机动交通的组织越发缺乏对人的关注，公共空间的可进入性降低，公共空间之间缺乏联系，导致其使用效率下降。

城市的生态绿网在不断演进，但成斑块式的模式，有待形成联网式的城市绿网体系。

本方案以城市公共空间的发展、通风廊道和绿色廊道走势为出发点，结合世博时及世博后的策划内容，对本场地的公共空间进行了定位和重构；带状绿色空间，不但增大绿化辐射面，而且引导围合了各有特色的公共空间场所，以此强调环境的自然生态性，强调减弱人与自然的互相冲突，利用环境的和谐增强人与人之间的沟通交流，创造绿色的城市公共开放空间。

With the development of city, the rapid growth of traffic volume and the organization of motor transport plus the relationship between people and public space are out of concern. The inaccessible of public space and the lack of contact among the public spaces lead to its decline in using efficiency.

Green Ecological Net of Urban is becoming a mottling-like pattern, as the city develops. It has the opportunity to net each other to form an urban green net system.

This project takes the urban public space's development, and the well ventilated corridor plus the green corridor as the starting point. using the plan content of when and after the Expo 2010, we restructure the public space; Belt-shaped green space, not only expands the green service surface, but also guids and encircles a characteristic public space, the project gives emphasis to the environment's natural ecology, weakens the conflict between the human and the natural, and make use of the harmonious environment to enhancement people's communication, in prder to create an open green public space.

绿色城市，绿色网络

## 经济技术指标

总用地面积： 23.4hm²
总建筑面积： 9.4hm²
容 积 率： 0.4
建筑密度： 20%
绿 地 率： 45%

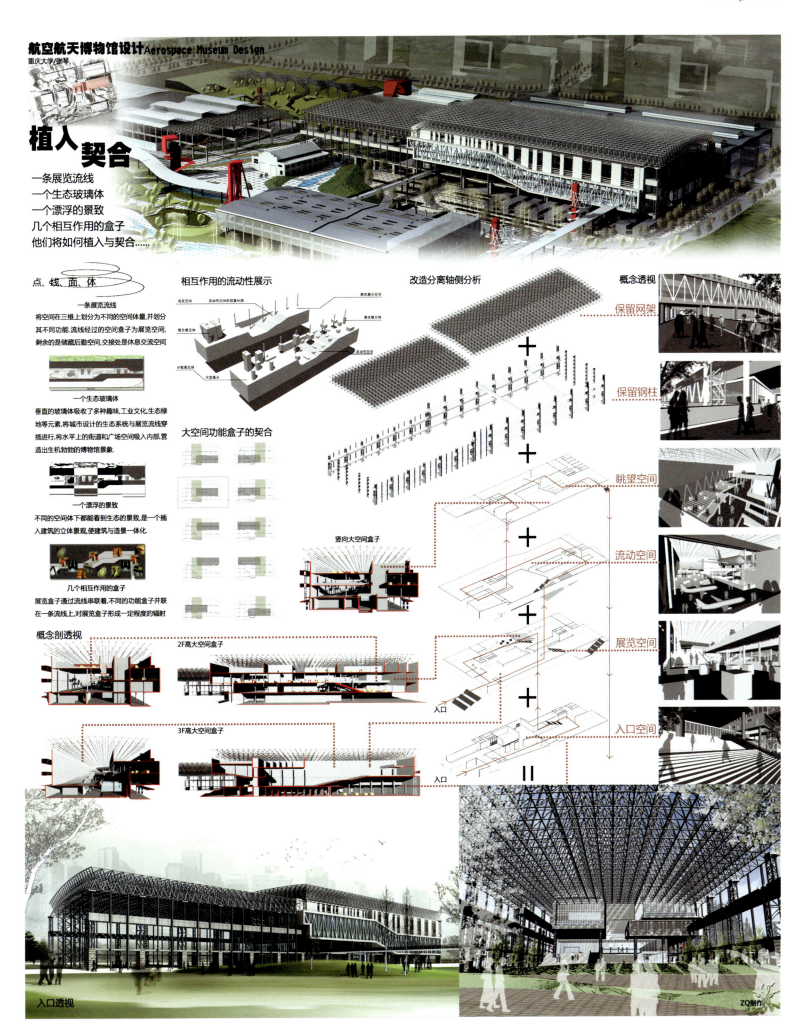

# ADAPTIVE REGENERATION OF PROCESSING WORKSHOP
## 2010 EXPO SHANGHAI JIANGNAN SHIPYARD

2,400M LEVEL PLAN

6,600M LEVEL PLAN

11,300M LEVEL PLAN

17,000M LEVEL PLAN

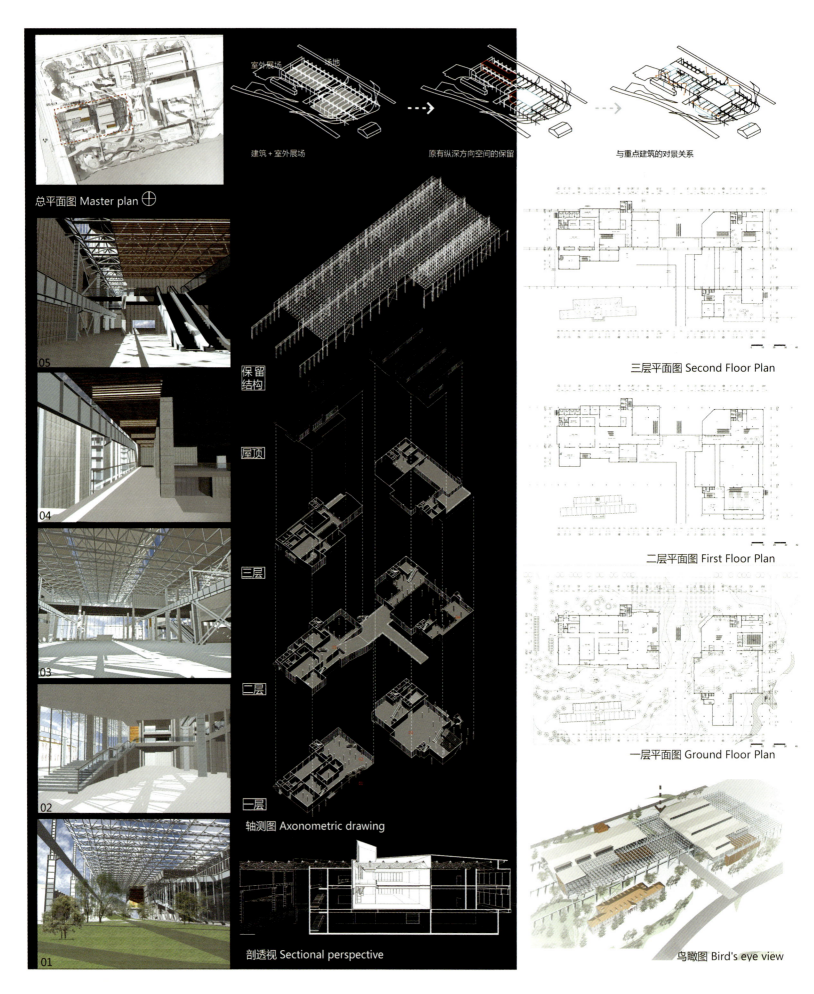

# 近现代工业博物馆

基地区位关系
zone bit of the site

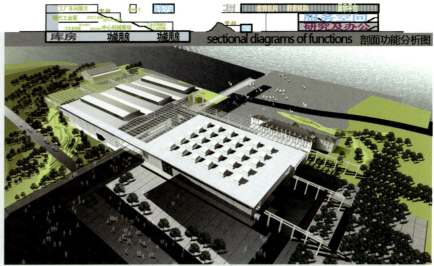

sectional diagrams of functions 剖面功能分析图

总体鸟瞰图
over-all bird-eye perspective

老建筑柱网及吊车
existing structure and runway girder

二层高架交换体系
over-head transformation terrace of the second floor

室内外大中厅示意图
diagrams of interior and exterior courtyards

新建的梁柱体系
new structures

最终成果
final existence

设计过程分析
analysis of design process

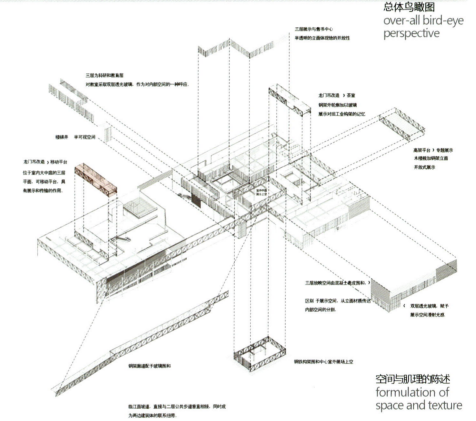

空间与肌理的陈述
formulation of space and texture

建筑位于二层步道及安检入口的交叉点上,如何安排好"穿越"及"上升"这两个意象很重要,游客在参观漫步中,同时游走到高架体系或穿越到江滨码头,保留工厂老建对人的记忆,运用厂房内的动力装置,形成可动空间,围绕室内外的大中庭空产生了螺旋上升的参观路径,在世博后的后续使用上考虑的是把东面作为科教展示中心,西面继续作为厂房的大空间展示(从剖面图上可见)。

入口透视图 entrance perspective

一层平面图 1st floor plan

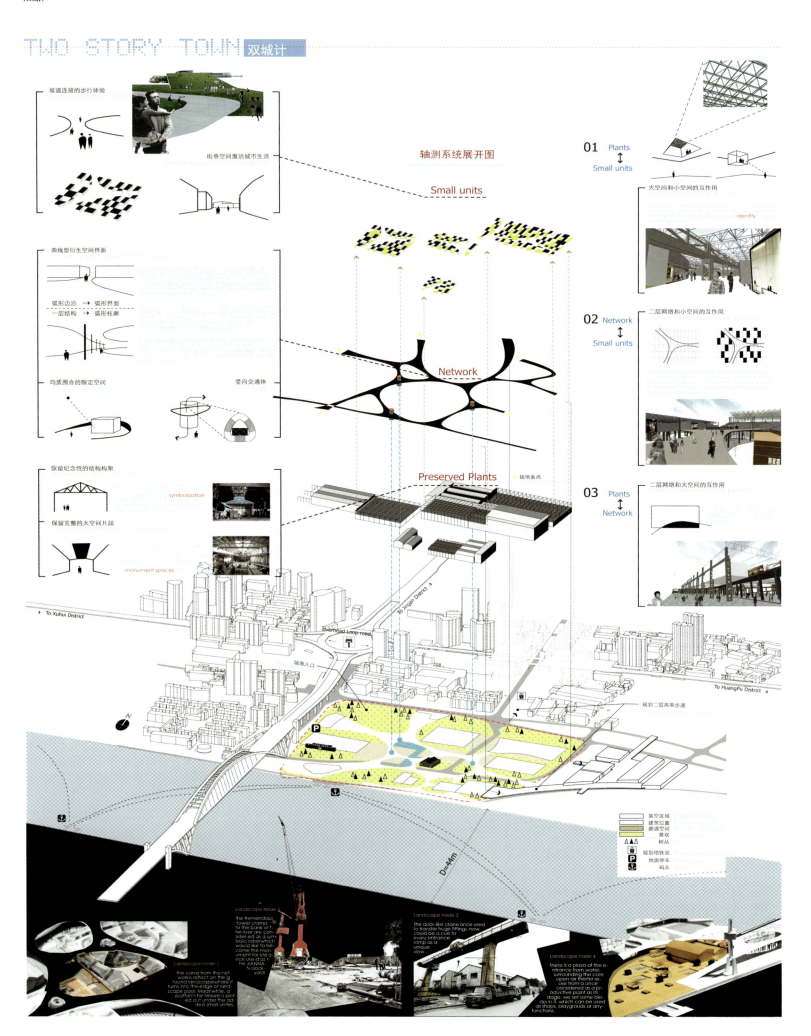

重庆大学 〉城市建筑的更新与再生--2010年上海世博会中国现代工业博物馆群设计 〉设计：**周雪峰 孙海霆 崔耀 王怡雯 韩文昌**
Chongqing University >Regeneration of Urban Architecture-Museum Complex of China Morden Industry for EXPO2010 Shanghai

# 双城计 Two Story Towns

身份重叠的故事 A Story of Overlap Identities

我们理解城市设计是一个动态的过程，它是一个在时间空间上规则和弹性的平衡。2010年上海世博会中国近代工业博物馆群项目面临的机遇和挑战同样巨大而复杂：它将位于上海核心区的一块2.3hm²的百年历史工业基地厂区——江南造船厂经过改造,作为世博会浦西主入口之一,迎接700万参观游客；并在世博之后，保持博物馆群的项目定位，融入城市生活。我们在建立了世博会和城市关系的理解之后，明确了关注后续利用，将地块融入城市公共生活的设计立场；由城市密度研究的视角入手，以高效的公共空间使用为基本设计目标，在工业遗迹保护和空间潜力挖掘之间取得平衡。在规划的条件下建立一个立体的连续流动的曲面步行系统，将场地所有的"身份"和项目以多维的方式无缝地联系在一起。在新和旧，大和小、上和下之间获得相互的身份识别和自明。并籍此激活一个空中的步行城市，将此空中城市的活力辐射到地面和空中。在尽可能保护现有建筑空间和历史记忆的情况下，提升空间使用效率，开发城市公共空间资源，有效提升整体城市公共生活品质。

We consider urban design as a dynamic process, and a balance reached between regulations and elasticities of time and space as well. The Modern Chinese Industrial museum complex project of 2010 Shanghai World Expo is faced with chances and challenges both tremendous and complex: it has transformed an 2.3hm² industrial base—Jiangnan Dockyard, at the cordial part of Shanghai, into one of the main entrance of Puxi part of the World Expo, meeting 7million visitors then; And after that, it will maintain its orientation and blend itself into urban life.After we get an acknowledgement of the relationship between the World Expo and city, we are clear that the design standpoint should emphasize on the subsequent usage and public life of city; Starting from the perspective of research on the urban density, we set efficient usage of public space as the basic design objective, and we aim at achieving balance between protection of industrial relics and exploration of spatial potent. Establish a three dimensional consistently-flowing curved surface pedestrian system, and connect all the identifications and projects through a multidimensional way. Obtain the mutual identification between new and old, big and small, up and down, and vitalize a pedestrian city in the air, and radiate the energy of this city to the earth and the sky. Try to improve the efficiency of space utility, explore the urban public spatial resource, and promote the overall public living standard of urban area efficiently, under the condition of protecting the existing architectural space and historical memory as much as possible.

We consider urban design as a dynamic process, and a balance reached between regulations and elasticities of time and space as well. Therefore, the objective of our urban design is set on achieving the temporal and spatial harmony between control and elasticity.

The Modern Chinese Industrial museum complex project of 2010 Shanghai World Expo is faced with chances and challenges both tremendous and complex: it has transformed an 2.3ha industrial base of factories-Jiangnan Dockyard, which is at the cordial part of Shanghai, into one of the main entrance of Puxi part of the World Expo, meeting 7million visitors then; And after the World Expo, it will maintain its orientation and blend itself into urban life.

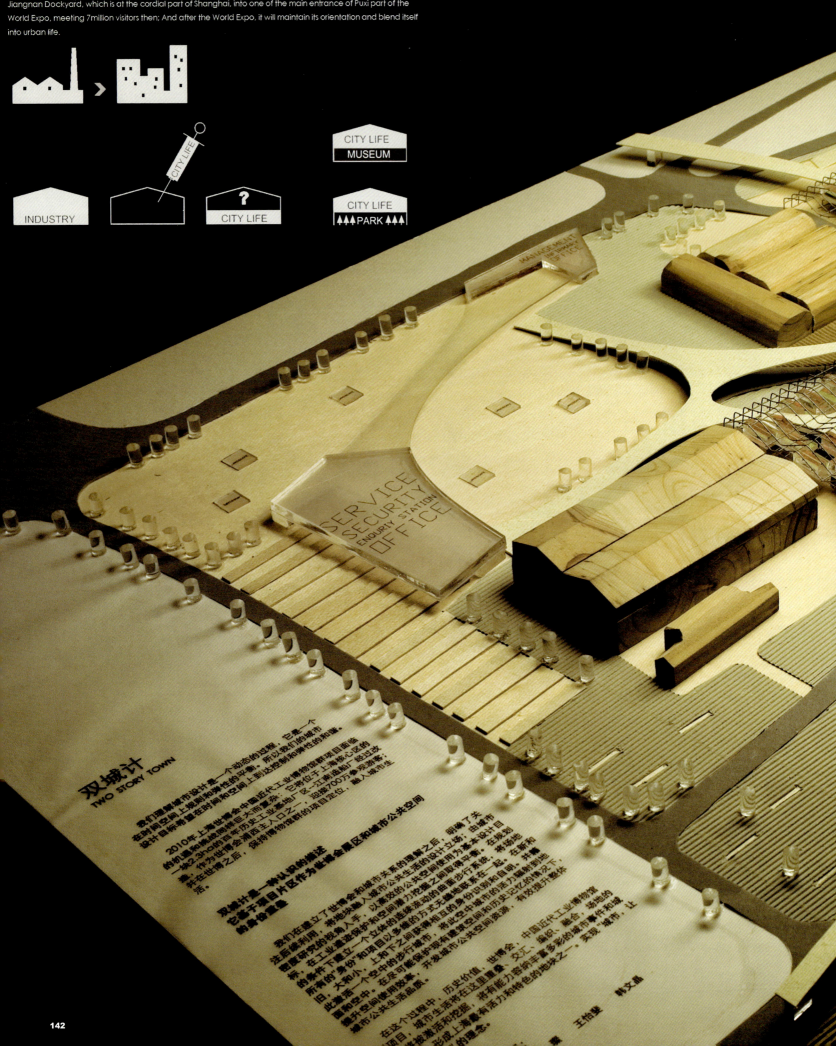

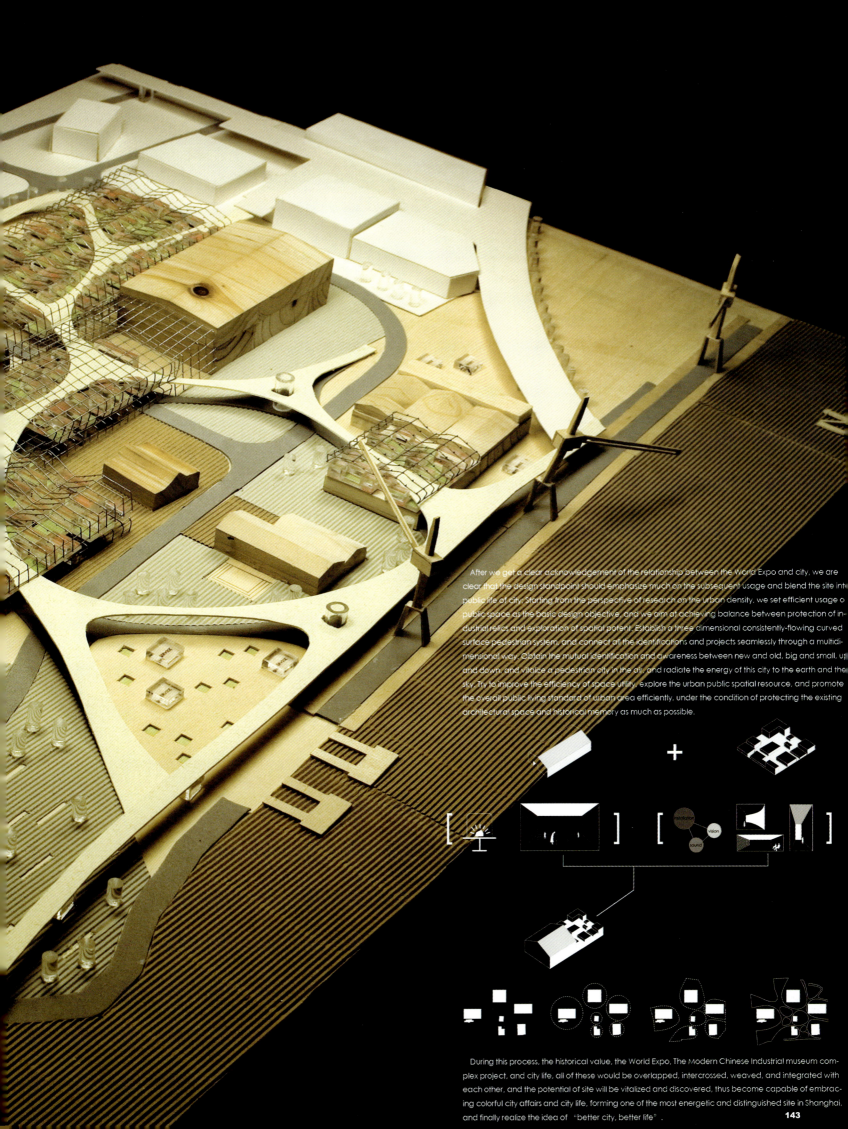

After we get a clear acknowledgement of the relationship between the World Expo and city, we are clear that the design standpoint should emphasize much on the subsequent usage and blend the site into public life of city. Starting from the perspective of research on the urban density, we set efficient usage of public space as the basic design objective, and we aim at achieving balance between protection of industrial relics and exploration of spatial potent. Establish a three dimensional consistently-flowing curved surface pedestrian system, and connect all the identifications and projects seamlessly through a multidimensional way. Obtain the mutual identification and awareness between new and old, big and small, up and down, and vitalize a pedestrian city in the air, and radiate the energy of this city to the earth and the sky. Try to improve the efficiency of space utility, explore the urban public spatial resource, and promote the overall public living standard of urban area efficiently, under the condition of protecting the existing architectural space and historical memory as much as possible.

During this process, the historical value, the World Expo, The Modern Chinese Industrial museum complex project, and city life, all of these would be overlapped, intercrossed, weaved, and integrated with each other, and the potential of site will be vitalized and discovered, thus become capable of embracing colorful city affairs and city life, forming one of the most energetic and distinguished site in Shanghai, and finally realize the idea of "better city, better life".

建筑设计导则
1 原则上将建筑分为改建区和拆建区
1.1 改建区为博物馆项目用地
1.2 拆建区为补充项目用地
1.3 改建区和拆建区分别设置设计导则

原状信息

| 西区装焊工厂 | 名称：加工车间 | 部件装焊工厂 | 管子制造工厂 | 船体联合车间 |
|---|---|---|---|---|
| 建造年代：1997~1999 | 建造年代：1984 | 建造年代：1998 | 建造年代：1983~1985 | 建造年代：20世纪90年代 |
| 主体结构：钢格构柱+四棱椎体网架 | 主体结构：混凝土排架 | 主体结构：钢柱+四棱椎体网架 | 主体结构：混凝土柱+排架/混凝土框架结构 | 主体结构：钢筋混凝土平面桁架 内部后加钢结构竖向支撑 |
| 外维护：工业用波纹彩围板 | 外围护清水红砖 | 外围护：下部分砖结构，上部分铝板维护 | 外围护：填充砖墙+粉刷外墙（白） | 外围护：砖结构 |
| 跨度：60m | 跨度：20m | 跨度：最大33m | 跨度：26m | 跨度：30m |

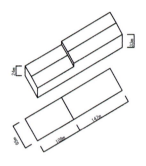  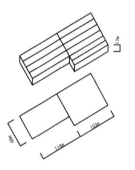 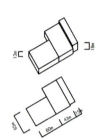 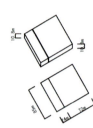

项目红线

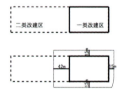  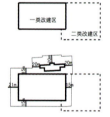  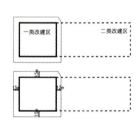

细则

图例

- 主开口面
- 可退不可突出
- 突出部分不超过相应深度和表面积的20%
- 与二层步道相接

一层平面

二层平面及以上

建筑限高

# 航海造船博物馆设计

重庆大学 周雪峰

## concept

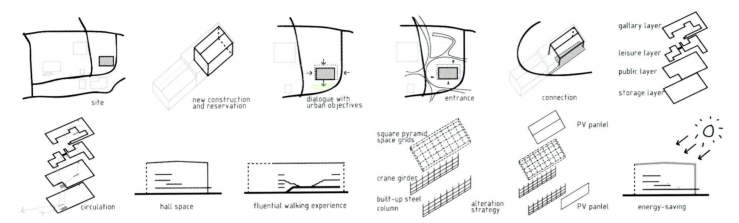

## view from entrance

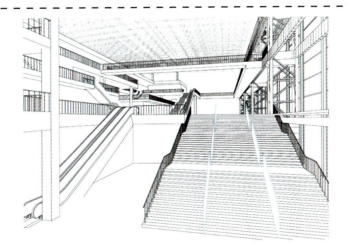

## drawing

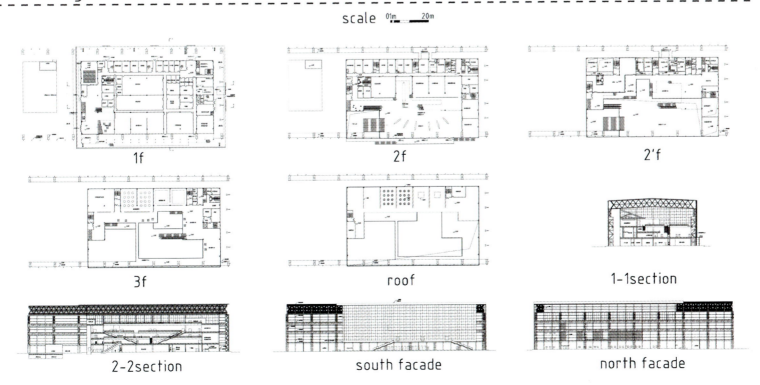

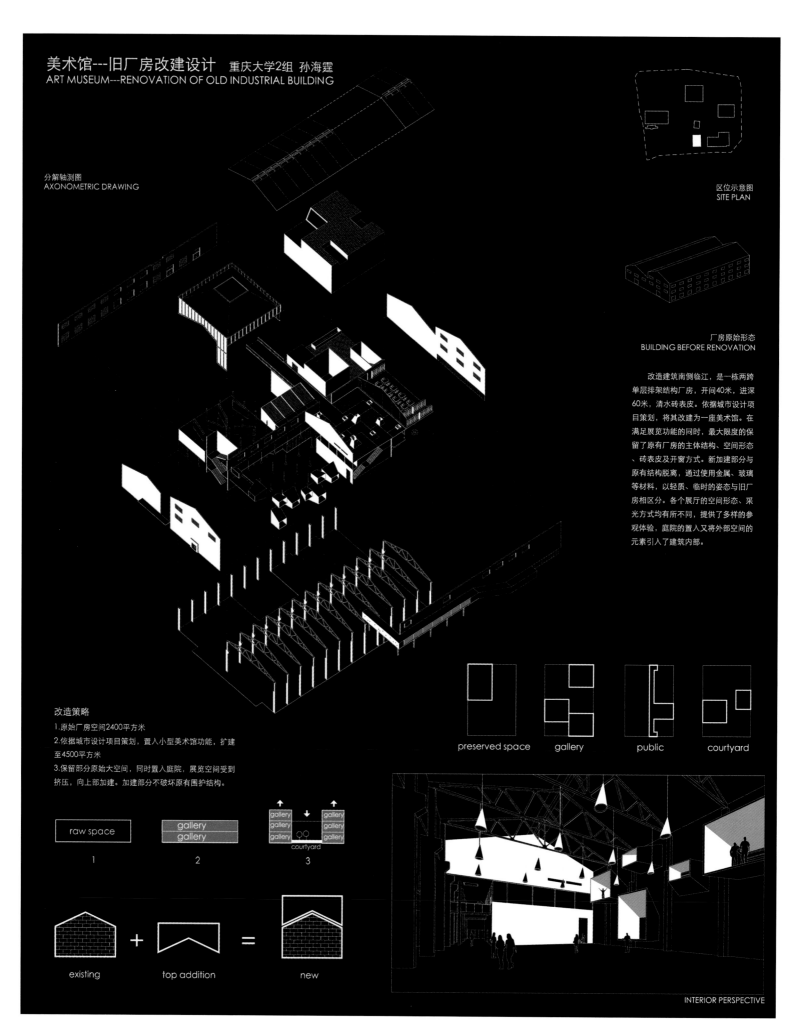

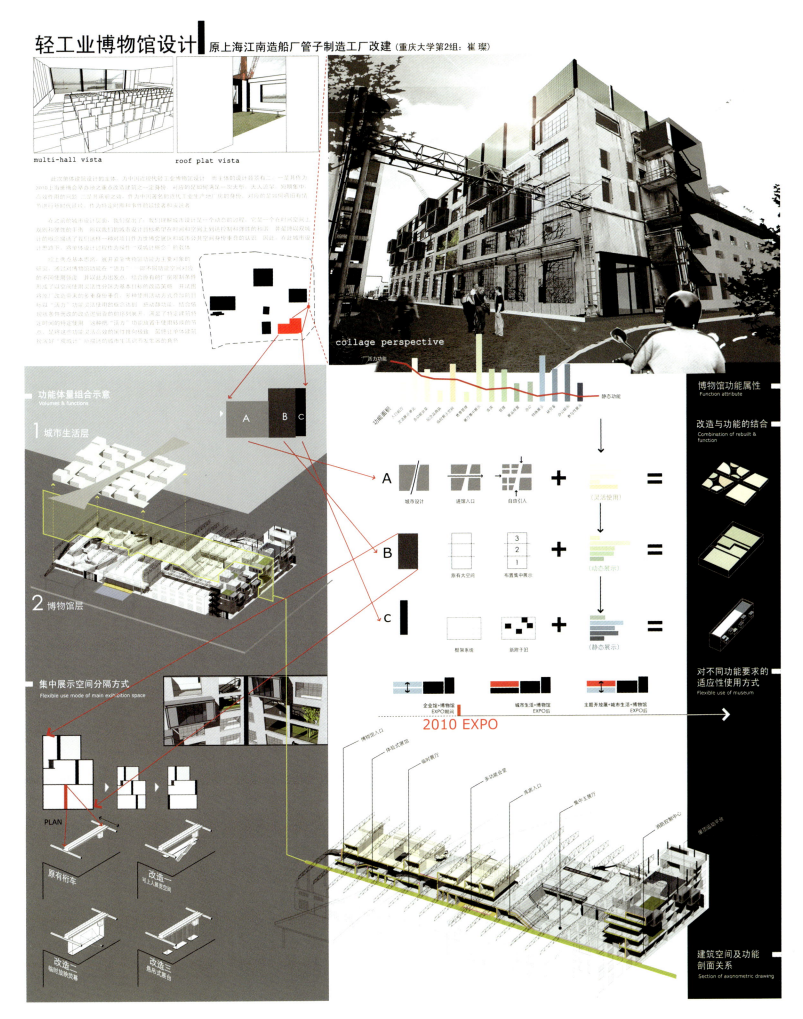

双城计

建筑退让 building setback

中国航空航天博物馆 CHINESE AEROSPACE MUSEUM

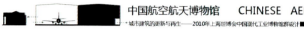

半室外展览　　　　　室内展览
semi-outdoor exhibition　　indoor exhibition

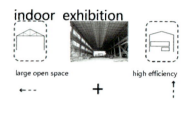
大型展品 large exhibits

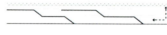
上升空间 climbing space

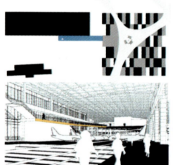

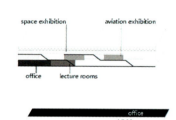
连接步道 connect to the ramp　　功能分布 function

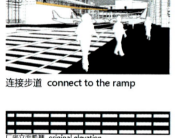
厂房立面肌理 original elevation

原有建筑立面的肌理对维持工厂记忆有一定作用，在半室外部分保留竖向的柱，使展品和博物馆旁的休息空间能相互渗透，让展览有更多的互动性。在正面室内展览部分保留了横向长窗，局部设置了通高的空间，一层退出一定商业空间，服务于城市。

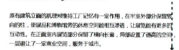
改造立面肌理 transformational elevation

改造立面肌理 transformational elevation
立面改造 elevation remodelling

## 工厂 → 博物馆

城市—世博—城市

在江南造船厂这样极具历史意义的地方进行改建，主要涉及到一个对老建筑的态度问题。这个设计项目是将原造船厂部件焊装工厂改造为中国航空航天博物馆。设计中首先遵循城市设计导则中的规定，保留了全部构架，并针对其紧邻的保护建筑海军司令部这一特点，退出一跨，结合航空航天的大型展品开辟半室外展场，同时结合场地的二层步道形成立体的参观空间。立面的改造上保留了原来厂房的虚实肌理，维持工厂记忆。在室内空间的安排上，首先保留了老工厂建筑结构形成的特有的极具纵深感的工业空间记忆；其次，针对其单层大空间的特点，决定在局部改造为多层空间，提高其使用效率。结合这两点，内部形成了一个在水平向和竖直向不断展开的爬升趋势，分别结合航空和航天发展了两条上升流线，既有分离又有循环。同时针对世博后人流变化的特点，将一层部分面向城市功能转换为商业，服务于城市。

148

【单体分解轴测】

# 近代工业资料馆设计
## ——船体联合车间改造设计

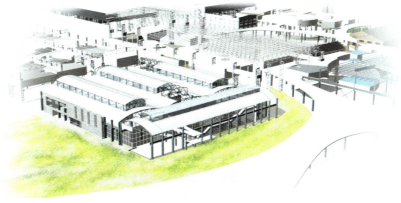

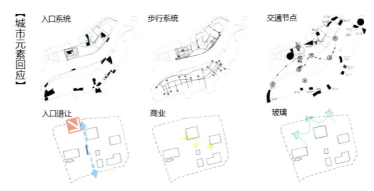

【城市元素回应】 入口系统 步行系统 交通节点

入口退让 商业 玻璃

【原空间特征 与 改造策略】

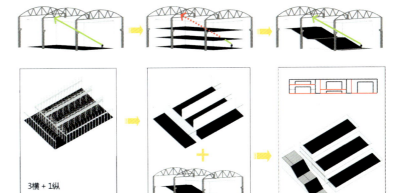

3横 + 1纵
钢筋混凝土平面桁架

【单体设计概念】 城市设计概念——过渡——单体设计概念

城市设计：大厂房+小房子　　保留单体大空间　　置入街道空间

大空间+街道空间

Wang Zhu,
Professor, doctoral tutor
Associate Dean of College of Civil Engineering and Architecture, head of the Department of Architecture, Zhejiang University
Main research directions: green architecture, regional architecture, urban landscape

**王竹**（1960.7—）
教授、博士生导师
浙江大学建筑工程学院副院长、建筑学系系主任
主要研究方向：绿色建筑、地域建筑、城市景观

Zhu Yuheng,
Associate professor, Ma Tutor
Assistant of the head of Department of Architect Zhejiang University
Main research directions: ecological living, traditic architecture protection, university planning

**朱宇恒**（1970.12—）
浙江大学建筑学系副教授、硕士生导师
建筑学系主任助理
主要研究方向：生态人居、传统建筑风貌保护、高校校园规划

**刘伯宇**
Liu Boyu

**孙翌**
Sun Yi

**刘蕾**
Liu L

浙江大学

| 张黎源 | 杜鹃 | 公贤彦 | 张安琪 | 孙佩文 | 陶俊 |
|---|---|---|---|---|---|
| Zhang Liyuan | Du Juan | Gong Xianyan | Zhang Anqi | Sun Peiwen | Tao Jun |

邓超
Deng Chao

祝马丽
Zhu Mali

MEMBERS: BOYU L, CHAO D, MALI ZH, YI S, LEI L.
ZHEJIANG UNIVERSITY. GROUP 1.

# 时之容器
# TIME CONTAINER

关于上海，侬想啥能 :)
WHAT DO U THINK OF SH

艺术展示
二手店，旧书摊

WIFI，留学，招聘
我什么都想要！！！

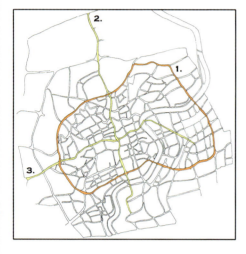

### MAIN CIRCULATION
1. ZHONG SHAN ROAD
2. THE NORTH-SOUTH ELEVATED ROAD
3. YAN AN ELEVATED ROAD

### PLACES FOR EXHIBITION
1. EVERBRIGHT CONVENTION & EXHIBITION CENTRE
2. SHANGHAI INTERNATIONAL EXHIBITION CENTRE
3. SHANGHAI MUSEUM & SHANGHAI GALLERY
4. INTERNATIONAL CONVENTION CENTRE
5. SHANGHAI NEW INTERNATIONAL EXPO CENTRE

### 城市交通主要流线
1. 中山路（上海内环线）
2. 南北高架路
3. 延安高架路

上海的"申"字交通体系承担了城市的主要交通负荷，南北高架和延安路高架在"十"字方向连接了城市的四方，同时串联了内外环线，内环线为城市的内部活动提供了最大的路面交通的便捷。

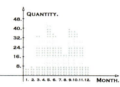

上图显示了上海展览场所的分布情况，我们可以发现在基地附近一定的半径内相关设施还是相当匮乏的，同时表明了此处发展战略的正确性。左图显示了上海计划在2008年举行的展会时间分布表，我们可以从中找寻展会的淡季和旺季。

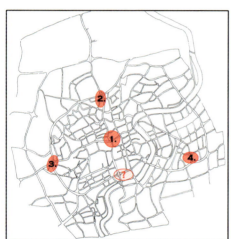

### CITY MAIN NODE
1. PEOPLE'S SQUARE
2. SHANGHAI RAILWAY STATION
3. ZHONGSHAN PARK
4. CENTURY PARK
?. OUR SITE

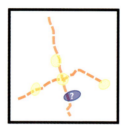

二元 VS 时间

二元 VS 功能

### 城市功能主要节点
1. 人民广场区域
2. 上海火车站区域
3. 中山公园区域
4. 世纪公园区域
5. 我们的场地

在"申"字体系的五个交点区域，形成了城市的五个主要综合节点。中间的人民广场无疑承担了最多的城市功能。在这些节点中有交通焦点，商业焦点，开放空间焦点等。我们的场地同时处于这一交点位置，我们希望能够提供场所的独特性，并且在最大的程度上为城市服务。

如果说一元是必须的，那么二元则是bonus，在现有基地的功能要求得到满足后，我们希望积极的引入第二元素来进行补充，让场所的生命力得以更好的延展。一元的工业博物馆是固定的，是为来自四面八方的人们服务的，那么作为第二元的城市市集则是活动的，功能可变的，更多的是为上海的城市生活服务的。由于它的可变性，所以能够在不同时间上取得多种可能性。我们在分析城市展览淡季旺季和进行了调查问卷后，拟定了一份可能的城市市集功能演变时间表，以更好的服务大众。

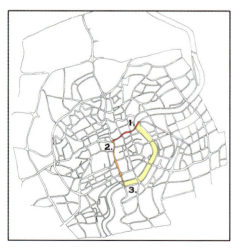

### TOURISM TRIANGLE
1. NANJING ROAD (E)
2. CHENGDU ROAD CHONGQING ROAD LUBAN ROAD
3. THE BUND

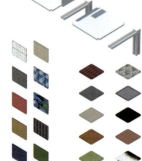

我们尝试运用完全可拆卸的构筑形式来满足对于不同功能的不同空间要求，从分解轴测我们可以清楚的感知其如何进行拼接。由于它的可变性，所以对于它的每一块"零件"来说，无论是作为立面还是作为底面顶面，她的材质都是可以根据具体要求而进行改变的。

### 城市旅游黄金三角
1. 南京路商业街
2. 南北向商业连接（成都路，重庆路，鲁班路）
3. 外滩

南京路商业步行街和外滩无疑是上海最吸引人的旅游景点。南京路体现了商业，外滩体现了文化，纵向的非延续界面更多的承担了交通功能。我们的场地处于这一三角形的一角，我们希望他也能够成为这黄金三角的一环，在今后更加闪光耀眼。

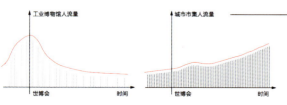

## CITY ANALYSIS
## 城市分析

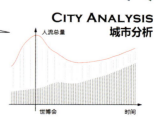

以时间为主轴，世博为节点，纵向测算人流密度。可以想见：在世博会期间，场地的使用者以游客为主；世博之后以周围的居民为主，辅以随集市而来的其他市民和零散游客，形成稳定的波动。

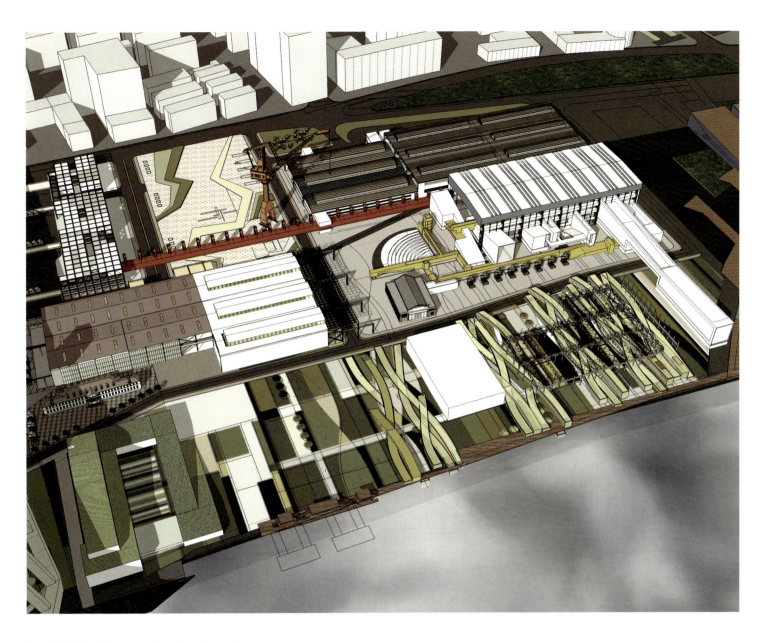

Zhejiang University > TIME CONTAINER

# 时之容器 Time Container

城市建筑的更新与再生——2010年上海世博会中国现代工业博物馆群设计

　　城市设计与2010EXPO规划和上海市城市规划相契合，世博园围栏线、二号船坞遗址公园、城市公交终端……都是我们设计注重的细节。西北角全新的"城市补丁"在形式上沿袭工业厂房的构成方法，多种空间组合可以适应城市不同的需求。如果说工业博物馆群是属于世界、中国，附加的第二元则属于上海、市民。城市设计的时间尺度可以以世纪为单位，我们力求城市地块得以自然和谐地更新。

　　历史建筑、巨构厂房旧有的空间、结构及材质等无法满足工业博物馆群的功能要求，我们在红线范围内用多种不同的手法尝试创新式的最大保留。遵循旧有结构模数的"加法"、弱化冷酷厂房的"减法"、新结构单体复制的"乘法"、仅保留骨架的"除法"－分别体现在工业馆群、集散广场和总馆、城市补丁和沿江地带的设计中。不同的风格承载不同时代的记忆。

　　地块功能二元叠加，功能更新自然过渡、场地记忆最大保留、巨构厂房解构再生。

时间的断面在此地呈现。

We do urban design of this site with consideration of larger scale plan--the 2010-EXPO design and the city plan of Shanghai, so that this site will be renewed in harmony with the city.
Traditional architecture, tremedous faory space, structure and fabric do not patch with what is required in museums.
With kinds of urban design innovation, the site memory and space character will be well remained, while various memories and figures of ares are brought in, thus sections of time emerge here.

关键词：

二元叠加　最大保留　自然过渡　解构再生

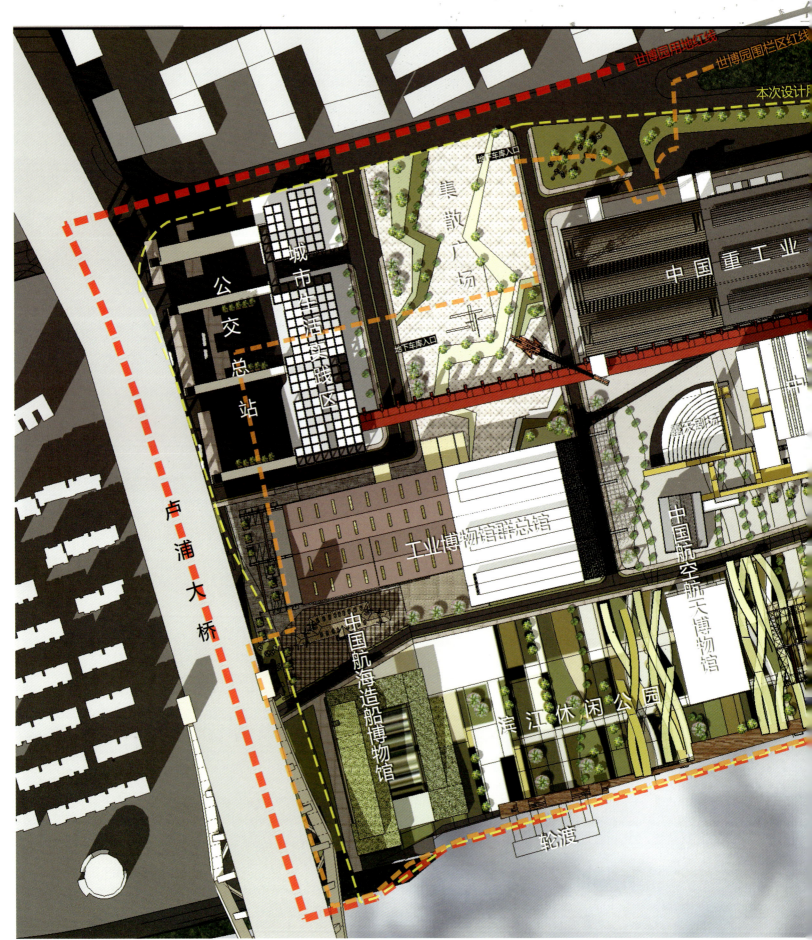

MEMBERS: BOYU L, CHAO D, MALI ZH, YI S, LEI L.
ZHEJIANG UNIVERSITY. GROUP 1.

时之容器

# TIME CONTAINER

Key Word:
+ Function   − Space
× Style   ÷ Construction

黄 浦 江

1 : 1000

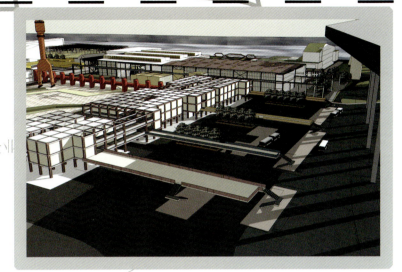

城市补丁

东北鸟瞰

东南鸟瞰

时之容器

# INDUSTRY MUSEUMS 2.0
## 工业博物馆群 2.0

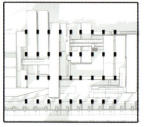 + 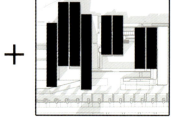 + 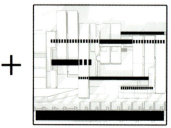 →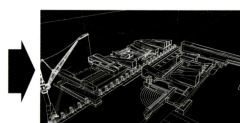

原有构架，构成博物馆建筑的基本网格语言。
The reserved struction makes up the base grid language of museums building.

加建建筑以南北向矩形为基本语言嵌入原有厂房的结构中。
The south-north rectangle make up the plan of adding building, and put into the reserved buillding.

东西向廊道连结起加建建筑，不同的廊宽形成不同的使用空间。
Adding buildings are connected by several aisles East-West, different widthes make different space.

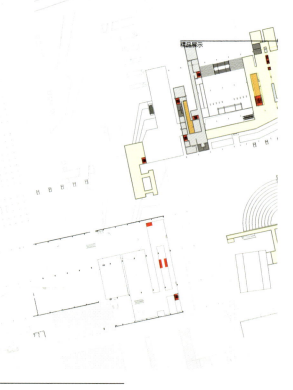

二层平面
SECOND FLOOR

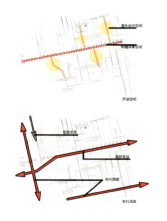

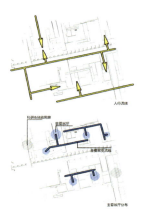

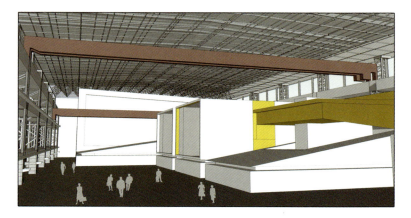

MEMBERS: BOYU L, CHAO D, MALI ZH, YI S, LEI L.
ZHEJIANG UNIVERSITY. GROUP 1.

# 时之容器
# TIME CONTAINER

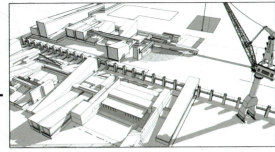

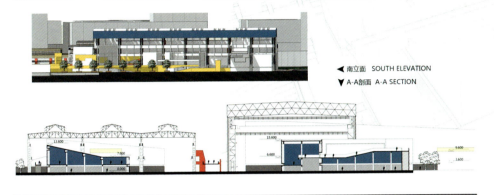

◄ 南立面 SOUTH ELEVATION
▼ A-A剖面 A-A SECTION

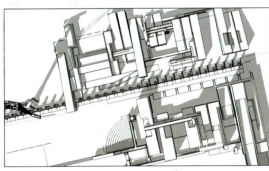

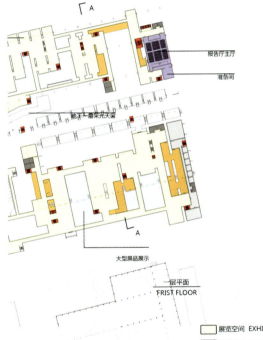

一层平面
FRIST FLOOR

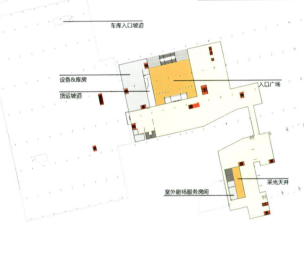

设计随笔：
1. 尽量依靠原有结构，实现保留厂房建筑价值的最大化。
2. 建筑单位的模数化，为日后的增建、改建提供可能性。
3. 建筑单位的可变性，通过建筑单位的增减，来实现空间尺度的变化。在满足不同尺度展品展览需要的同时，也提供了参观者不同的空间感受。
4. 上下自由穿越的连廊，将竖向空间连结成一个整体，在多维角度创造空间的趣味性。

地下一层平面
UNDERGROUND FLOOR

| | | | | |
|---|---|---|---|---|
| 展览空间 EXHIBIT SPACE | 垂直交通 VERTICAL TRAFFIC | 报告厅 AUDITORIUM | | 盥洗室 TOILET |
| 科研用房 HOUSE FOR SEARCHING | 室外平台 TERRACE OUTSIDE | 室外展区 SHOWING OUTSIDE | 地下车库 UNDERGROUND PARKING |

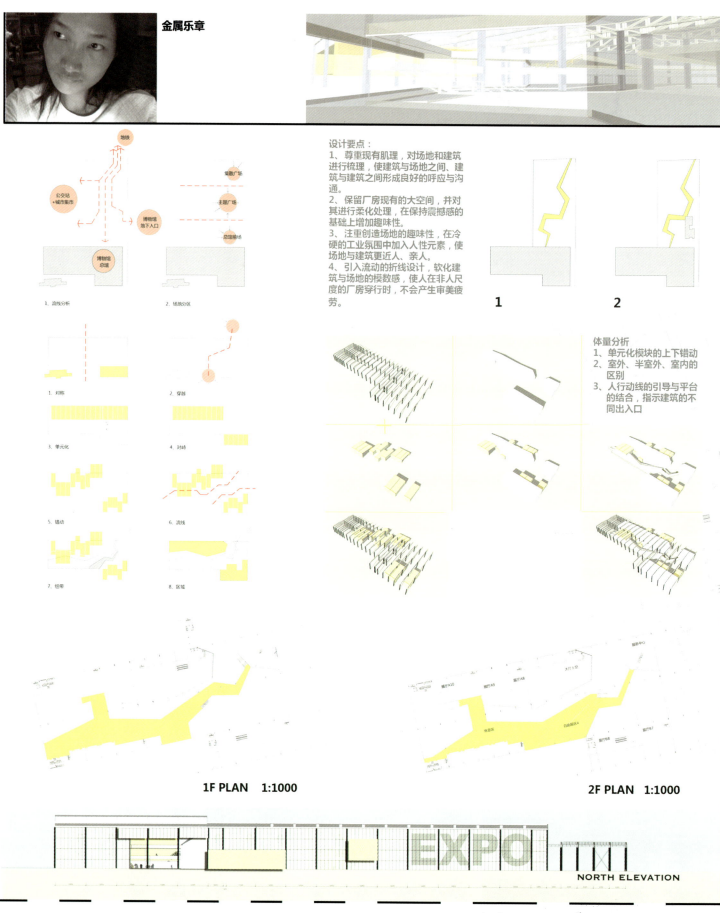

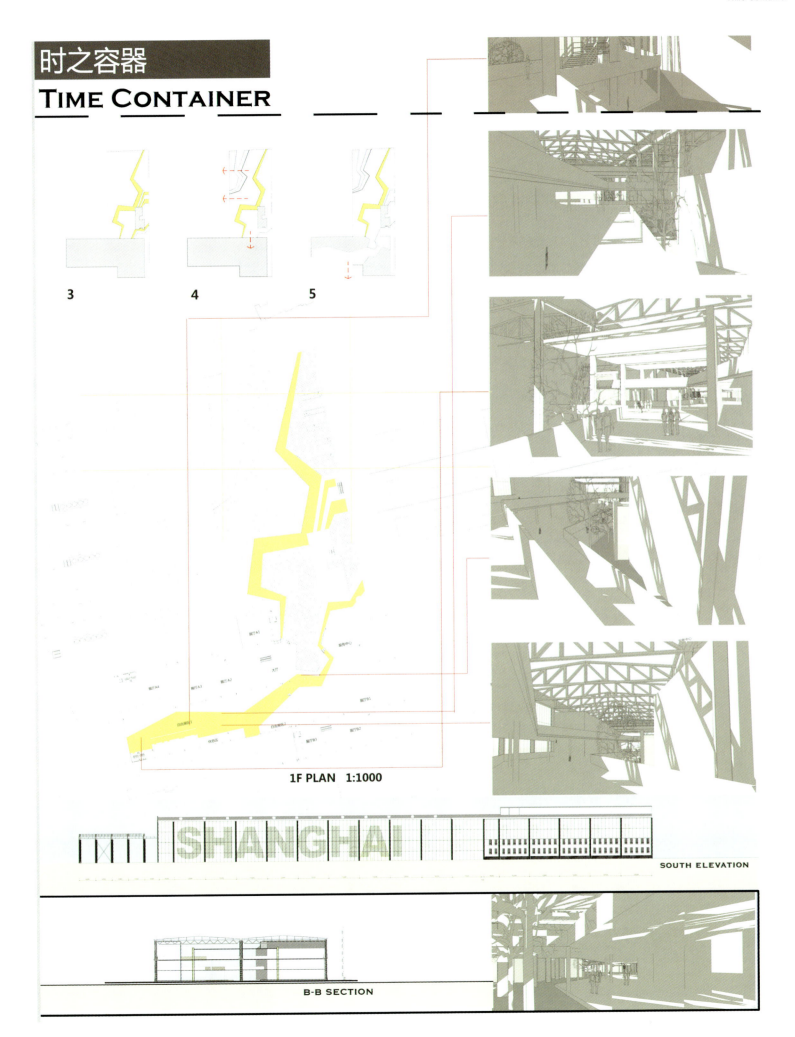

时之容器

城市补丁 2.0

------ 第二元．设计

### 关于停车场的设计
### ABOUT THE PARKING AREA

位于整个基地西北角的公交车客运中心需要配备十五条公交线路的停车及上下客的日常运营，同时在这个设计中，城市市集作为第二元，在服务于城市的同时吸引着一定数量的人流，作为一个重要的基地入口，为了让人流和车流分离，我们选择让行人下车后直接登上二层进入城市市集，避免与一层的公交车流线相冲突。为了保证一定的停车位，在城市市集的一层我们也预设了停车位，这样在白天，底层架空可以进行临时摊位的商业及其它功能的运营，而在晚上让位于停车位。

### 关于城市市集的设计
### ABOUT THE CITY PATCH

城市市集，为了满足它不同时期功能的变化，我们运用了可以完全拆卸的构筑形式。这里我们选择三种相对典型的平面形式来阐述我们的概念。这样的可变体系，可以营造不同的空间属性，可以呈现不同的建筑材质，可以创建不同的虚实感受和围合体验。对于城市市集的运营，我们希望政府可以根据上海城市的生活相关的指标和数据，在不同时期，根据功能和性质，租赁给不同的运营机构，他们可以是个人也可以是组织，对于空间的营造需要有专人负责统一筹控。

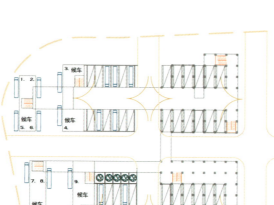

停车场及城市市集一层平面
PARKING AREA & CITY PATCH 1ST PLANE

### 关于综合展示的空间
### PLANE FOR THE EXHIBITION

1. 大空间 & 围合空间
   适用于不同性质的综合展示
2. 虚空间直接步入实空间
3. 卫生间结合楼梯间布置

### 关于人才市场的空间
### PLANE FOR THE JOB HUNTING

1. 大空间 结合 小空间
   适用于不同功能的综合运用
2. 虚空间过渡灰空间步入实空间
3. 大空间服务于报告会
   小空间服务于咨询投放简历
4. 卫生间结合楼梯间布置

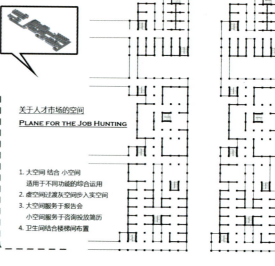

### 关于跳蚤市场的空间
### PLANE FOR THE MARKET

1. 小空间 & 半围合空间
   适用于租赁给个人的摊位
2. 虚空间直接步入灰空间
3. 卫生间结合楼梯间布置

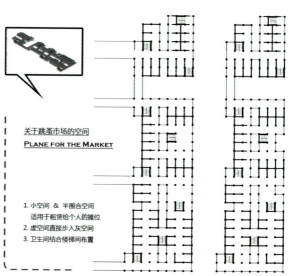

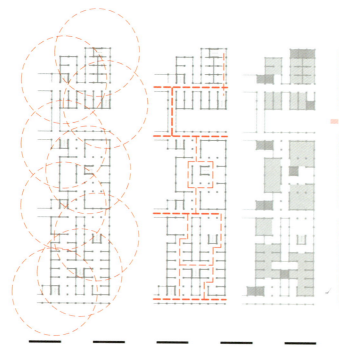

楼梯服务半径分析　　步行流线分析　　空间虚实分析　　轴线分区分析
ANA OF THE SERVING RADIUM　ANA OF THE PEDS' CIRCULATION　ANA OF THE VOIDS　ANA OF THE AXIS & FIELDS

# TIME CONTAINER
时之容器

# 时之容器
# TIME CONTAINER

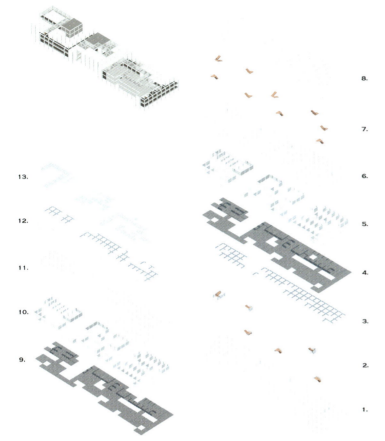

1. COLUMN OF 1ST FLOOR
2. CORE LINKING 1ST FLOOR & 2ND FLOOR
3. BEAM OF THE TOP OF 1ST FLOOR
4. CEILING OF 1ST FLOOR
5. WALL OF 2ND FLOOR
6. COLUMN OF 2ND FLOOR
7. CORE LINKING 2ND FLOOR & 3RD FLOOR
8. BEAM OF THE TOP OF 2ND FLOOR
9. CEILING OF 2ND FLOOR
10. WALL OF 3RD FLOOR
11. COLUMN OF 3RD FLOOR
12. BEAM OF THE TOP OF 3RD FLOOR
13. CEILING OF THE 3RD FLOOR

A. MAIN PERSPECTIVE
B. OUT-DETAIL PERSPECTIVE
C. CORNER PERSPECTIVE
D. IN-YARD PERSPECTIVE

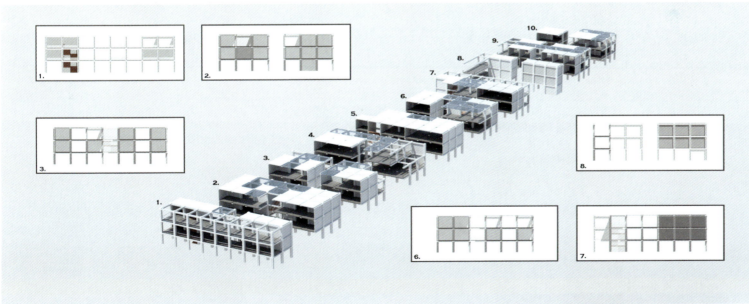

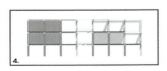
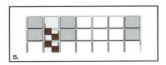

## CITY PATCH 2.0
### 城市补丁 2.0

IT'S DENGCHAO
ARPHITECTURE@HOTMAIL.COM

# TIME CONTAINER
## 时之容器

延续肌理 对话历史 景观编织

**Concept 1 Evolutionary texture**
策略一：延续肌理

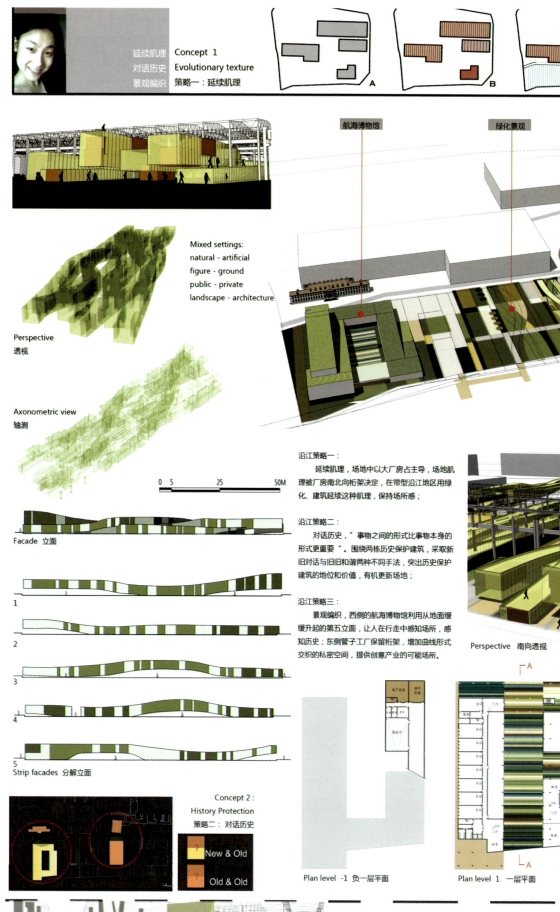

Mixed settings:
natural - artificial
figure - ground
public - private
landscape - architecture

Perspective 透视

Axonometric view 轴测

Bird view 鸟瞰

If the city is a landscape, buildings are mountains.
A ground as a building, a surface as volume.

Perspective 南向透视

Facade 立面

1
2
3
4
5
Strip facades 分解立面

**沿江策略一：**
延续肌理，场地中以大厂房占主导，场地肌理被厂房南北向桁架决定，在带型沿江地区用绿化、建筑延续这种肌理，保持场所感；

**沿江策略二：**
对话历史，"事物之间的形式比事物本身的形式更重要"。围绕两栋历史保护建筑，采取新旧对话与旧旧和谐两种不同手法，突出历史保护建筑的地位与价值，有机更新场地；

**沿江策略三：**
景观编织，西侧的航海博物馆利用从地面缓缓升起的第五立面，让人在行走中感知场所，感知历史；东侧管子工厂保留桁架，增加曲线形式交织的私密空间，提供创意产业的可能场所。

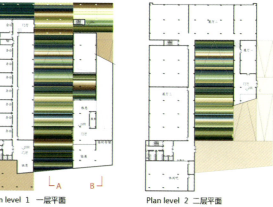

Concept 2:
History Protection
策略二：对话历史

New & Old
Old & Old

Plan level -1 负一层平面
Plan level 1 一层平面
Plan level 2 二层平面

# 时之容器

# TIME CONTAINER

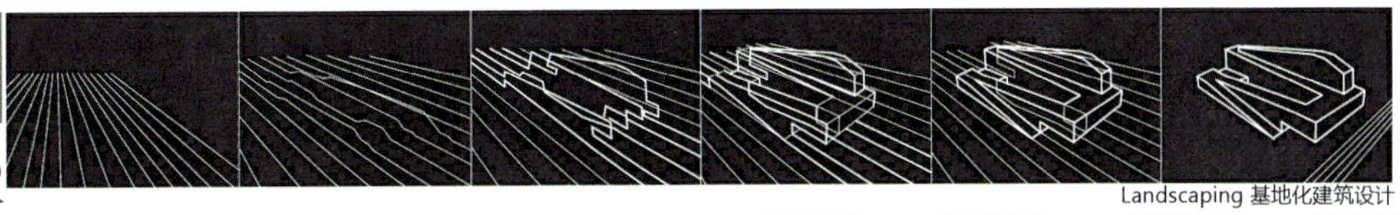

Landscaping 基地化建筑设计

创意产业/服务

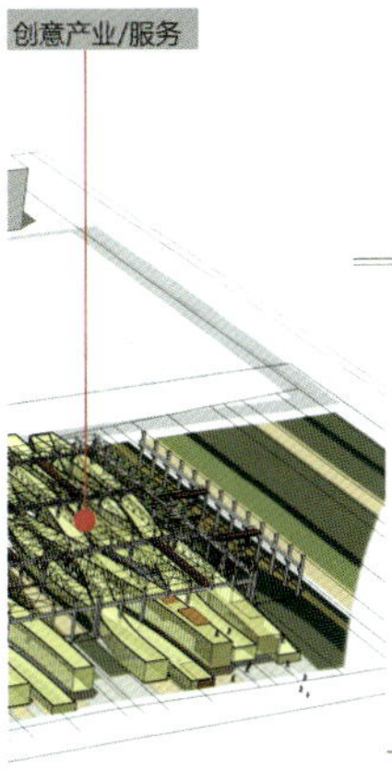

0 5 25 50M

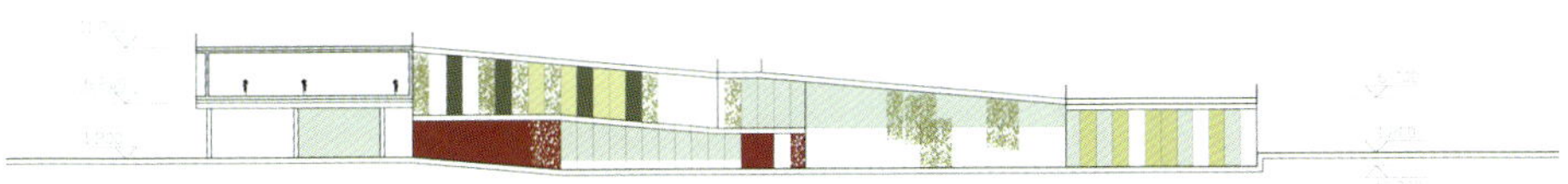

A-A剖面 Section courtyard

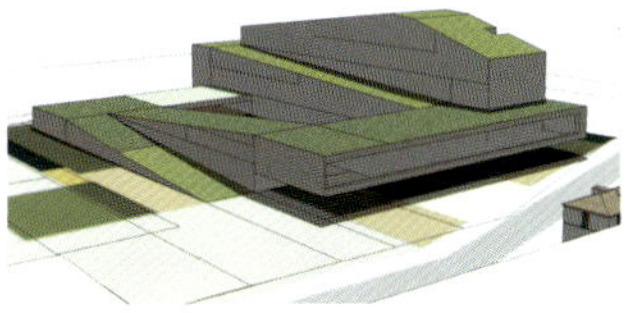

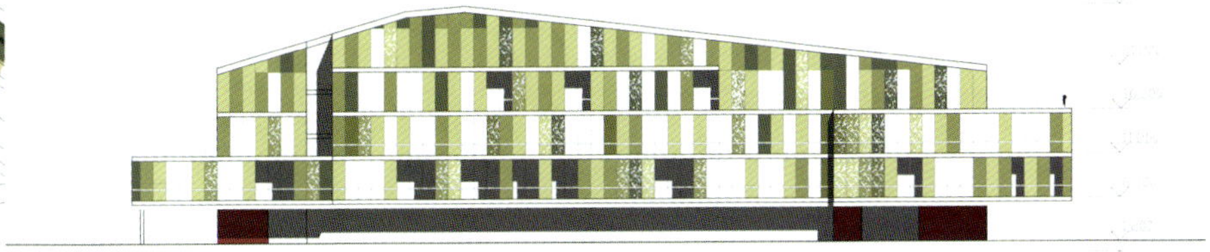

西立面 West elevation

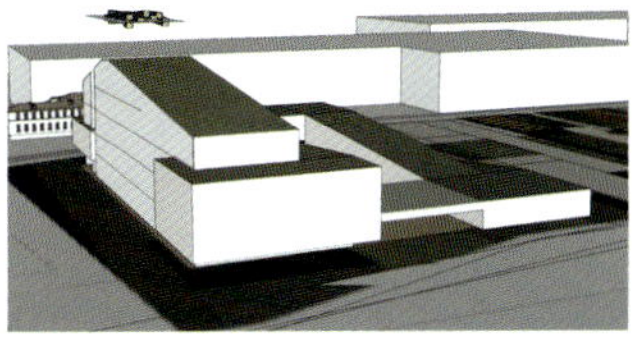

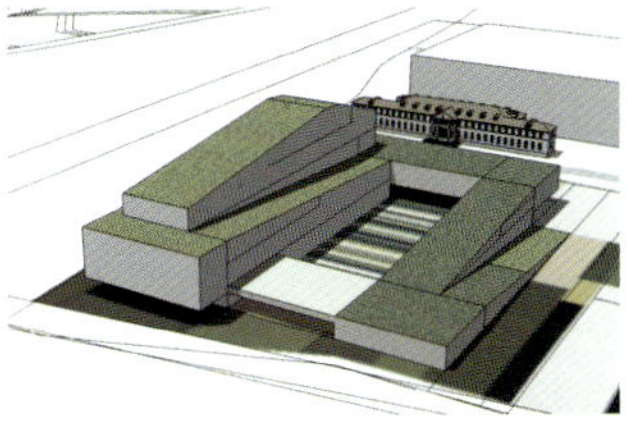

自然与人工，图与底，
公共与私密，景观与建筑

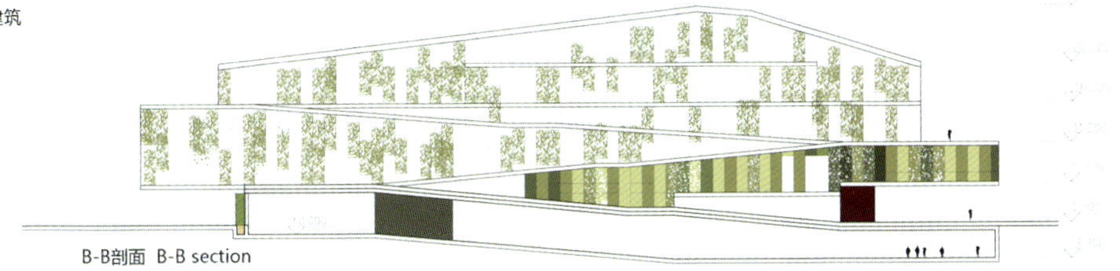

B-B剖面 B-B section

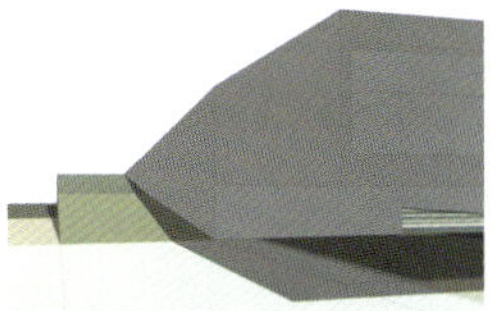

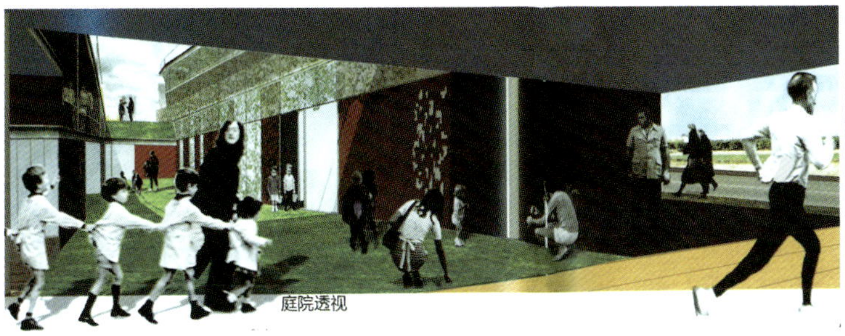

庭院透视

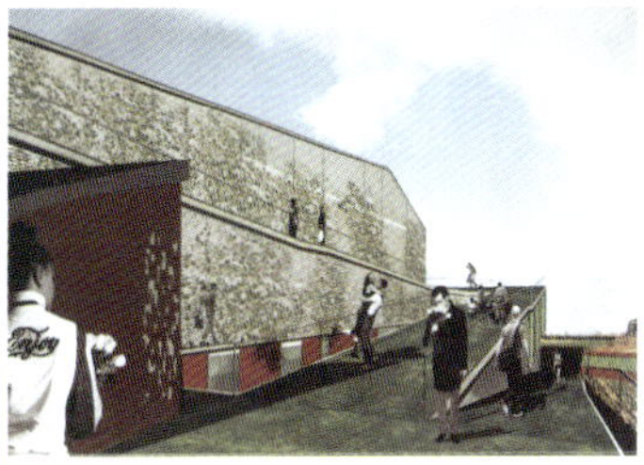

屋顶透视

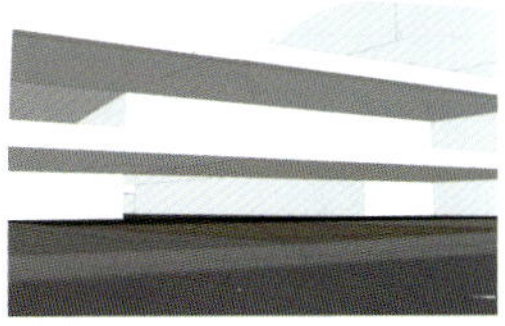

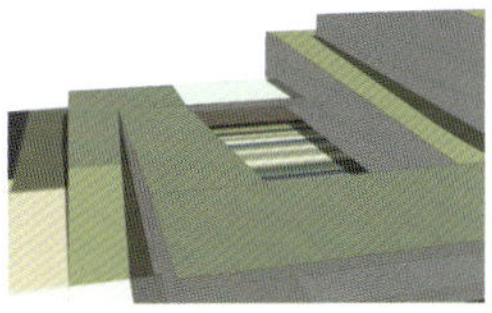

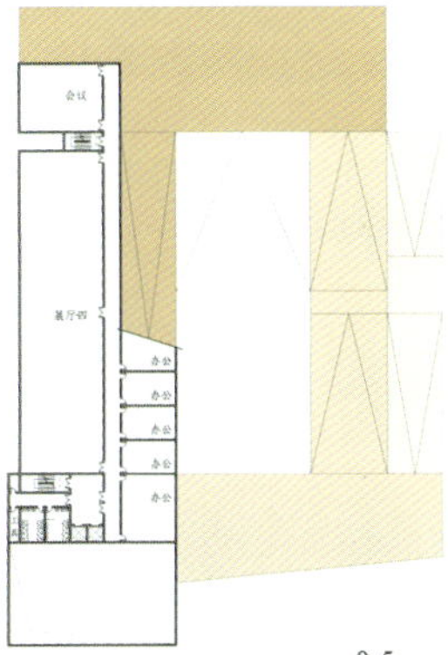

Plan level 3 三层平面

0 5 25 50M

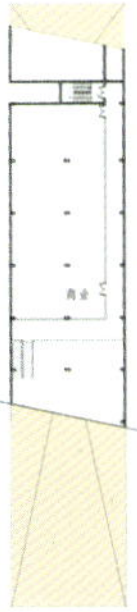

Plan level 5

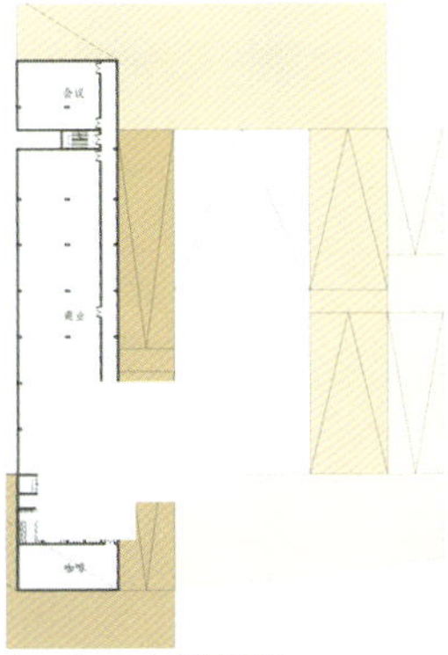

Plan level 4 四层平面

红砖楼的保留与历史建筑飞机库形成良好的对话关系，空间上南北贯通室内外剧场；东西串接景观；该楼的保留和改建使飞机库变为一个重要的人流节点，有效加强其中心地位。

Plan level 1

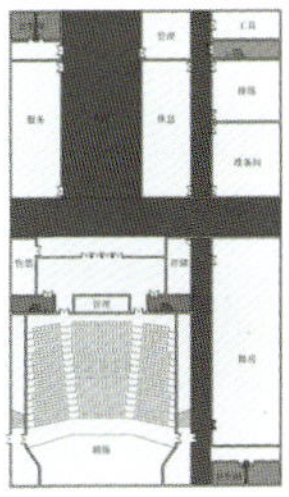

Plan level 2

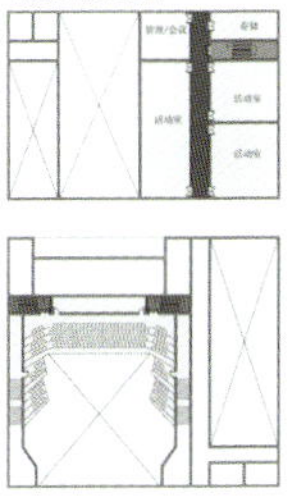

Space research
空间构成分析

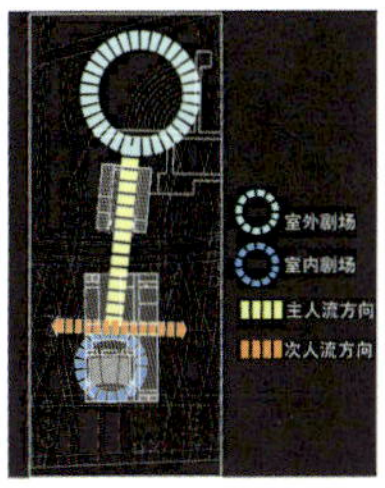


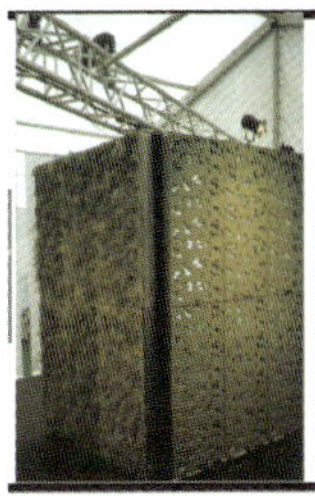

建筑材料
绿色的建筑材料营造与景观一体的航海博物馆

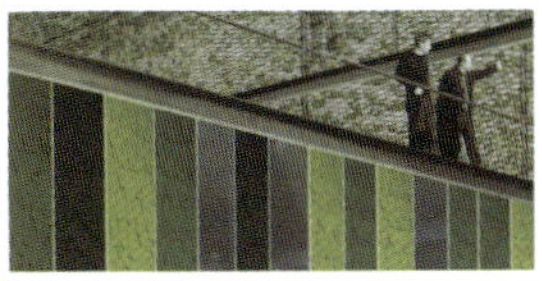

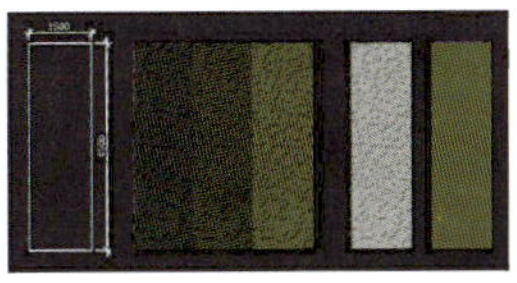

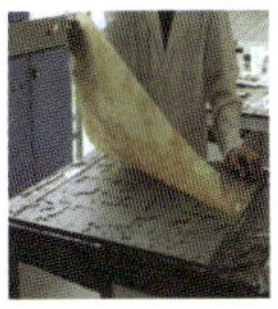

## WATERFRONT EVOLUTION
沿江策略

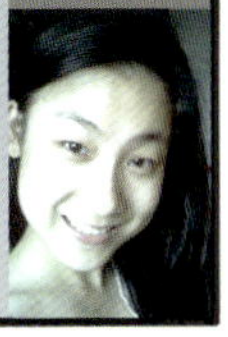

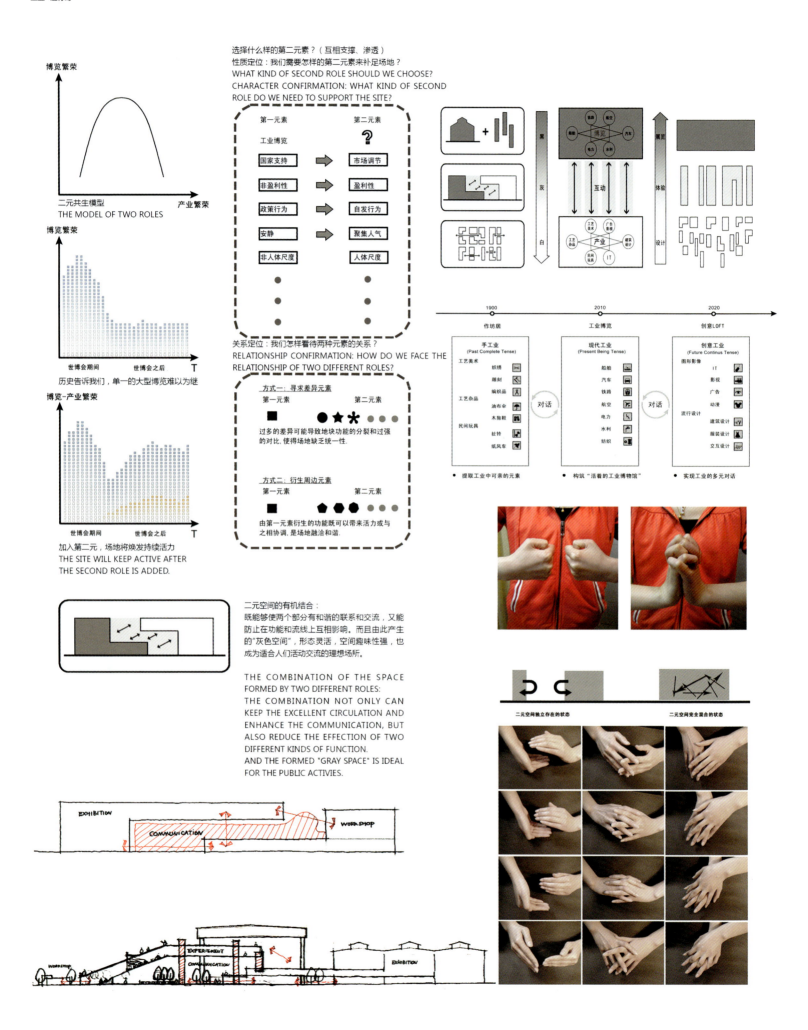

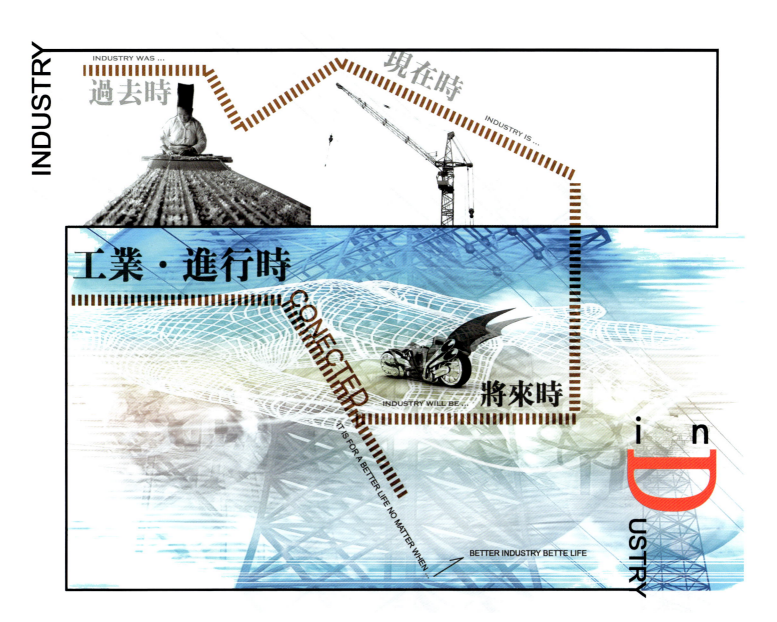

Zhejiang University >BETTER INDUSTRY, BETTER LIFE

# 工业 · 进行时　Better Industry, Better Life

上海世博园江南造船厂地块更新设计　RENOVATION OF THE JIANGNAN DOCKYARD FOR THE EXPO IN SHANGHAI

单一的博览产业难以独自生存，因此我们提出二元共生的概念，将一种确定的与工业博览同质的活力产业引入场地与其相辅相生，共同促进地块繁荣。而对于这个第二元素的定位，我们通过对工业的回溯和前瞻，确定为整个工业的历史发展及未来发展的过程，将历史的工业印迹和无限未来工业让生活更美好的各种可能性带入场地，使人们真正了解工业无所不在，推动着我们生活方方面面的本质。我们通过收纳工业过去式——手工业和工业的未来时——现代工业实践区，将固定参与人群引入场地，激活场地，支撑工业的现在时——工业博览。于是，大尺度的厂房空间通过人体尺度的匠人聚落——手工作坊和工业体验区从厂房顶部伸向江边的条状肌理得以丰富，并通过灰色空间扣成一个有张有弛，紧密宜人的整体。于此同时，人们与工业相结合的各种生存体现，生活状态和无限创意将形成一种互动的鲜活的工业展示，张显地场活力。由此，我们得以通过这一城市重要节点传承历史文脉，实现工业这一历史轴线的再现，实现死的工业博物馆向融入无限生活的鲜活的工业博物馆的转变，真正实现工业进行时。

Single expo industry is hard to survive. We propose the concept of dual symbiotic and introduce a new energetic industry into the area to support the former. As to defining the new industry, we look both back and into the future and decide that it should be the whole process of the development of the industry, including the past and the unknown future. By doing so, we intend to keep the memory of industry and bring to the area every possibility of happiness that industry can contribute to our life. Then people may truly understand that industry is everywhere promoting our society. By introducing the past tense of industry–handicrafts and the future tense of industry-new products experience center, we will introduce people to regularly participate in the area to activate it in order to support the present tense of industry–expo industry. The large scale provided by the factory would be enriched by the human scale that small architectures bring us, and they form an entity together with a connecting space. Meanwhile, the experience combining people and industry, the living style with industry and infinitude ideas would be a new kind of expo to revive the area. By doing this, we would be able to inherit the culture in this important spot, represent the axis of the industry history, transform the dead museum into one full of vigor and life, and truly achieve the present tense of the total industry.

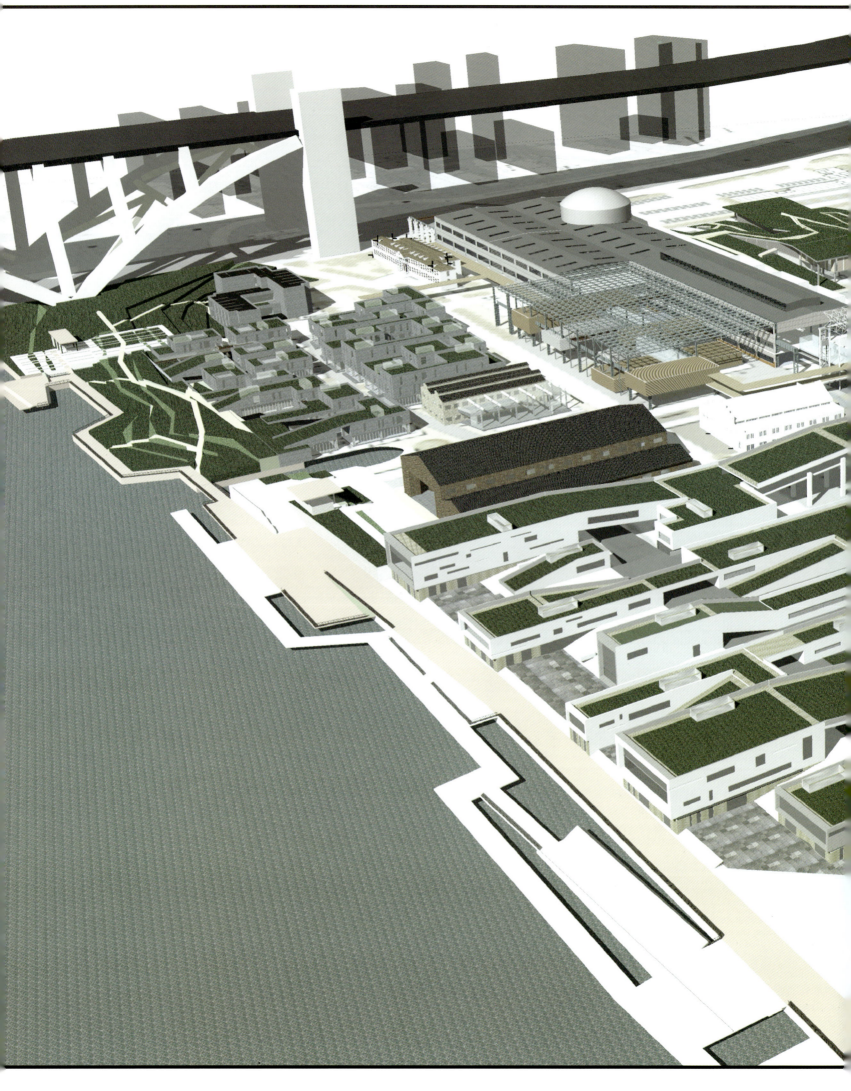

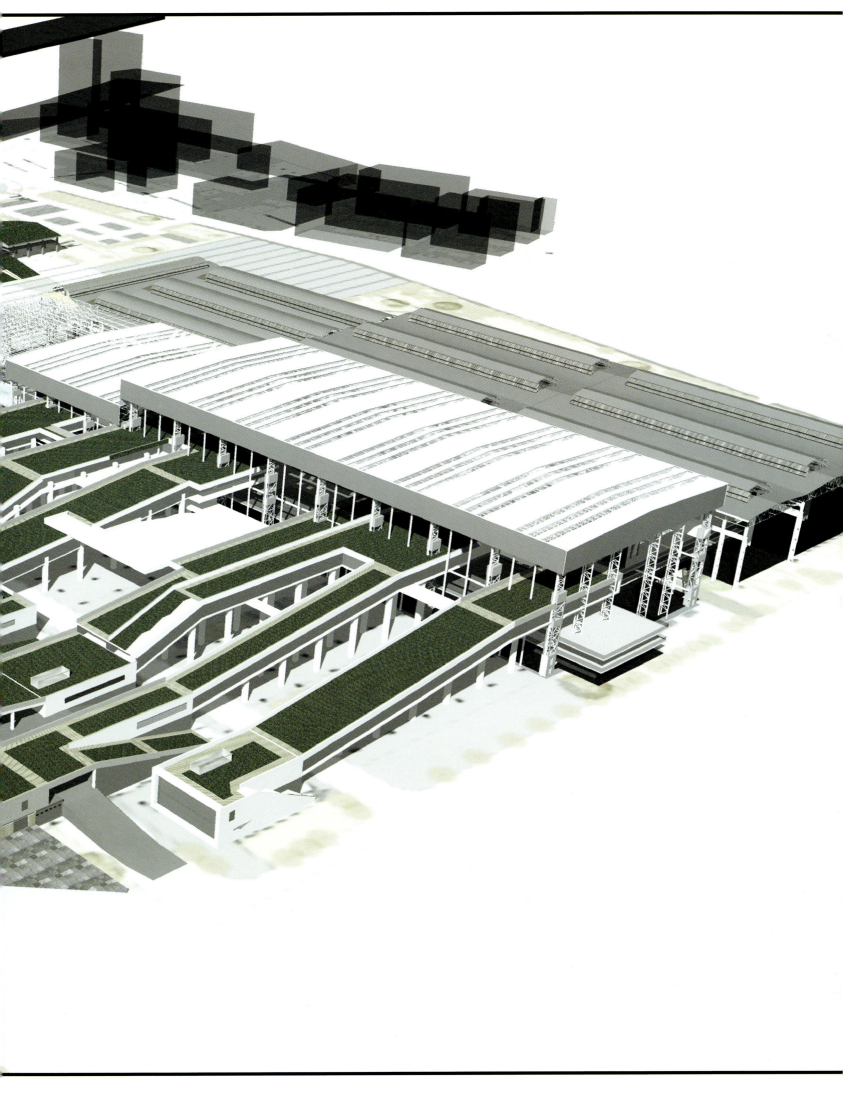

工业·进行时

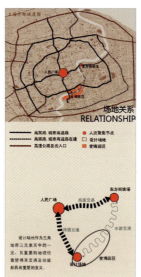
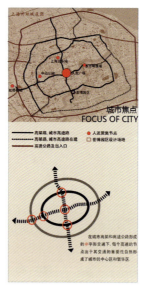

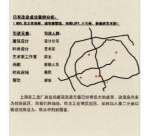
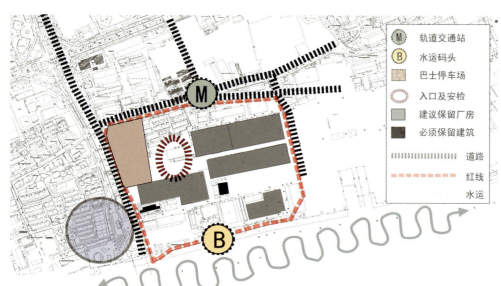

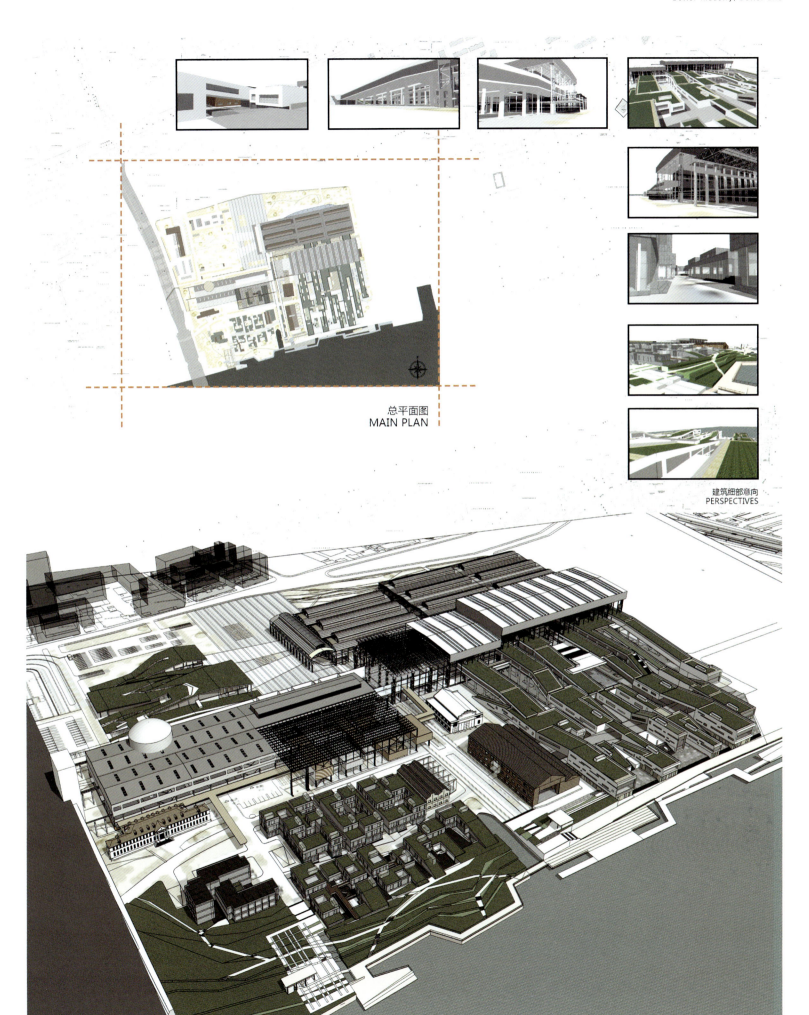

总平面图
MAIN PLAN

建筑细部意向
PERSPECTIVES

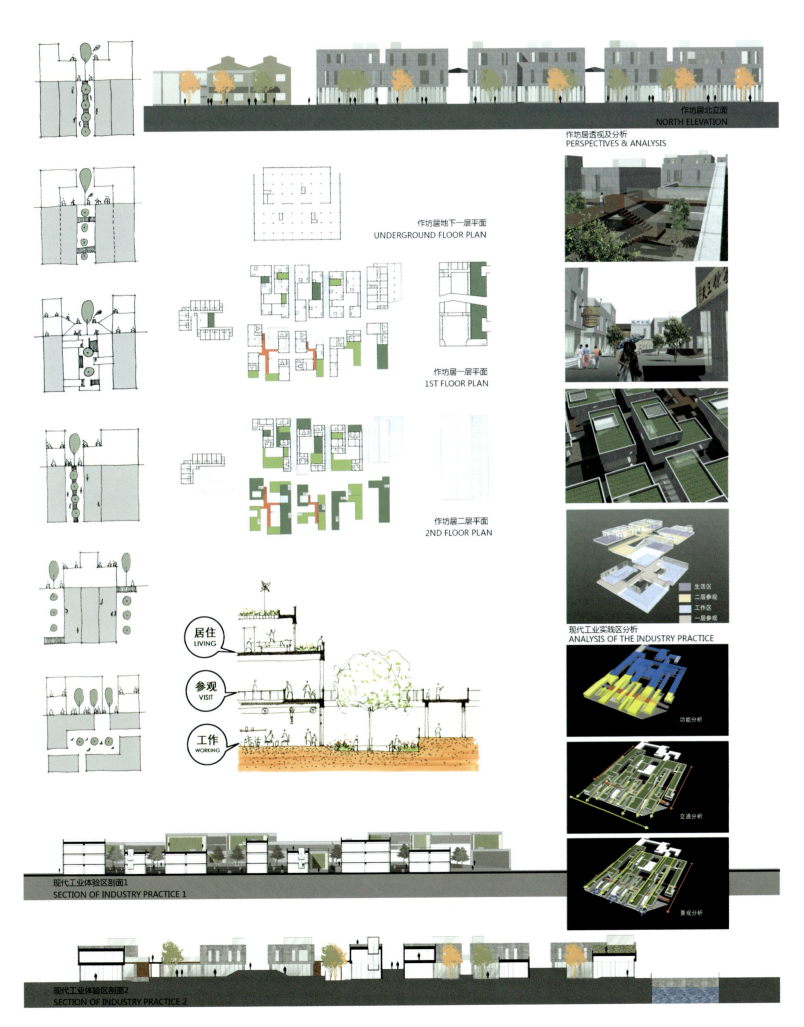

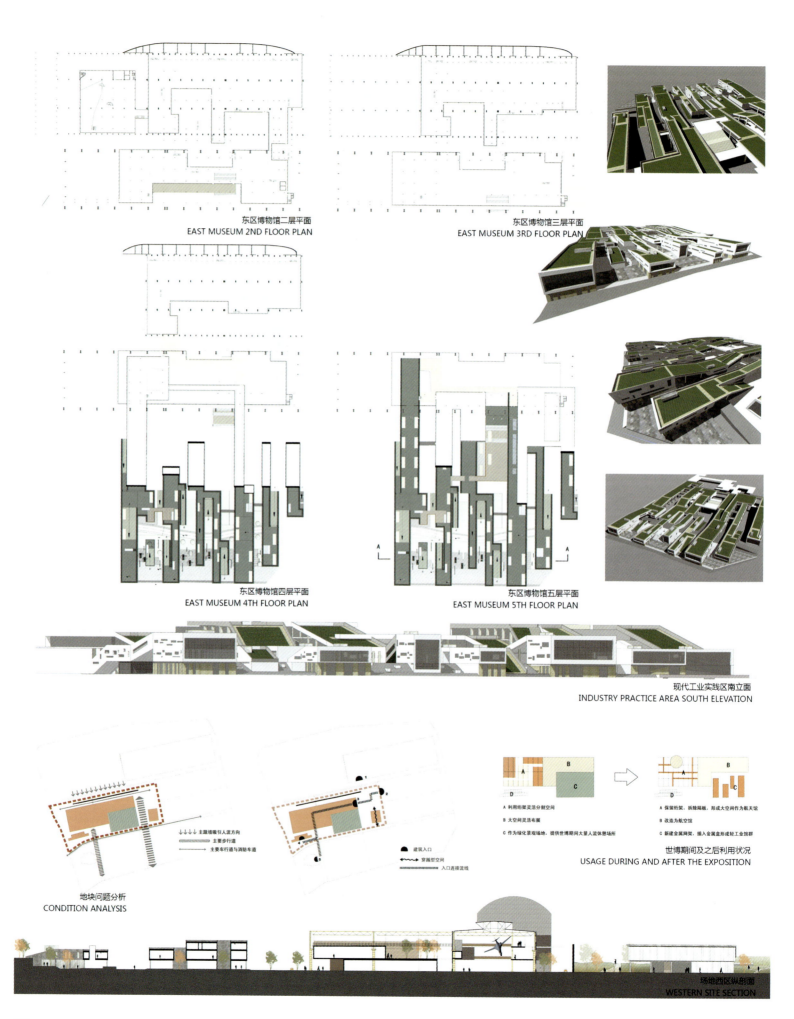

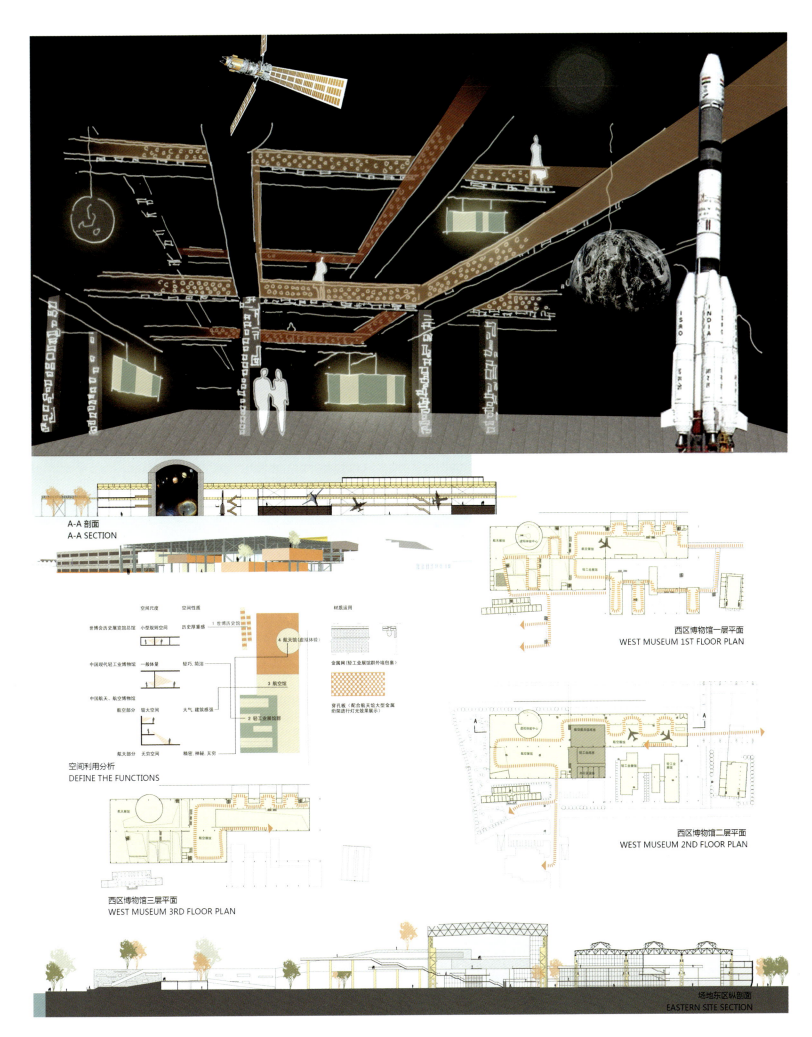

教师组成员

吕品晶　傅祎　周宇舫　刘彤昊　王铁　吴晓敏

设计小组成员

曹卿　张玉婷　袁野　杨剑雷　段敏　马超　苏迪

中央美术学院
China Central Academy of Fine Arts

生活让城市更美好

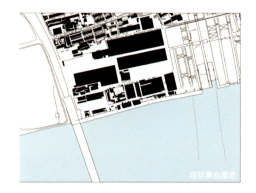
现状黑白图底

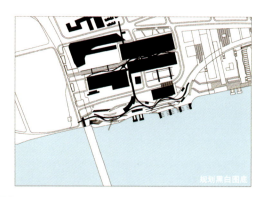
规划黑白图底

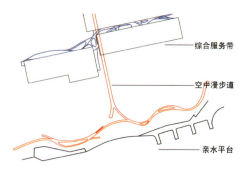
综合服务带
空中漫步道
亲水平台

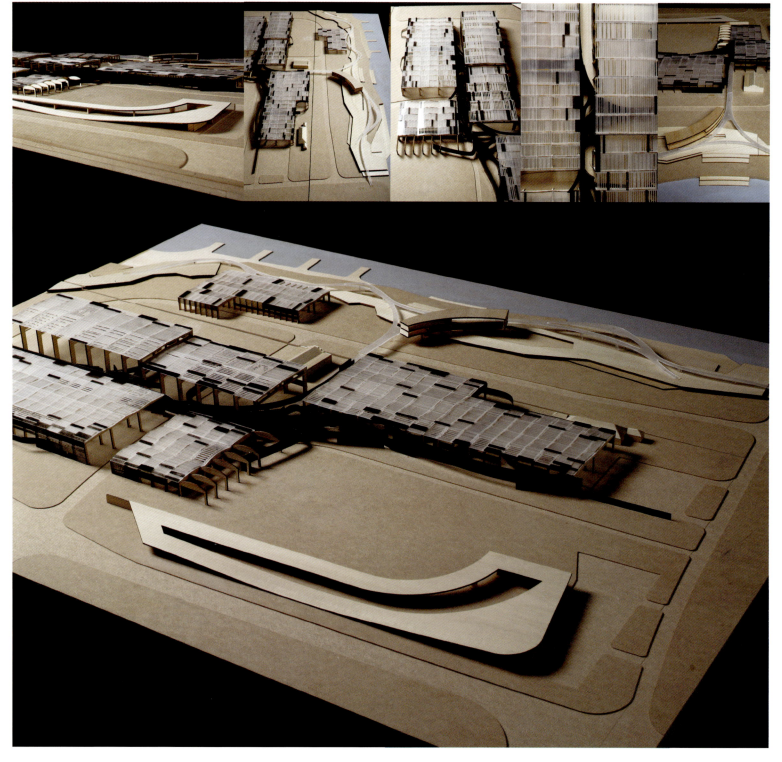

**中央美术学院** 〉江南造船厂城市设计与单体建筑改造 〉设计：曹卿 杨剑雷 张玉婷 袁野 段敏 马超 苏迪
Central Academy Of Fine Arts 〉Renovation Of Industrial Building

# 生活让城市更美好 Better Life Better City

　　世博会提出了BETTER CITY BETTER LIFE的口号，我们则提出了BETTER LIFE BETTER CITY，将这个口号反过来了，意在世博会之后如何用城市生活来延续整个地段的活力，通过对世博会理念和江南造船厂改造策略的深入理解，和对人们需要的体会得到任何旧厂区的改造都是为了使人们的生活更美好，人们作为城市的主体，也只有从人们的需求和生活出发才能使这种改造更有意义。Life的策略：（1）由"孤立"到"个性"。保留个性十足的工业厂房，充分利用滨水地带的自然优势，构建一个个性十足的区域。（2）由"封闭"到"延伸"。新区域将不会被孤立，在城市网络引入的同时将商业，旅游，文化活动伸入至此区域。（3）由"固定的形式"到"灵活的片断"。区域内原有的大空间为偶然事件提供了无限可能，一系列充满吸引力的功能空间，以混合的方式交叠在一起。（4）由"大尺度"到"景观"。位于滨水区域的绿色长廊为市民提供了休闲的空间。（5）由"设备"到"娱乐"。例如龙门吊等保留下来的工业设备可以再利用作为体验设备。（6）由"劳动"到"生动"。引入生活中人们需要的功能，满足不同人的需求。（7）由"工业"到"互动"。打破封闭的区域，形成完整的城市生活界面。使区域以一个开放的姿态与人们进行互动。

　　Expo took the "Better city, Better life" for its slogan. However, we reverse the slogan that is "Better life, Better city! The focus of our design is how to use the citizen's life to active and keep the vigor of the area. By in-depth understanding the idea of the Expo 2010 and the transformation strategy of the Jiang nan shipyard, we did knew that the aim of the transformation of the old factory is to make a better life for the people, especially in this area. As the main body of the city. So it is more meaningful to transform from a old state to a renovated state only from the people's need.The strategy of the "LIFE":"Isolation---identity"Identity reserve the plant which have the personality, the original landscape is another national advantage. "Enclosure"-"extension" As a open area it can stretch its boundary to the city's grid. Commerce junketing cultural activity can be offered in this place. "Fixed form"- flexibility fragment" The original big space offer the chances to various possibilities, a series of space which full of charming can be blended. "Large scale"— "landscape"The corridor located in the waterfront offering the citizens a variety of leisure spaces. "Equipment"—"entertainment"The devices such as cranes which were kept can re-used as experienced devices."Labor"-"live"The introduction of the needed functions such as commerce and cultural activities to satisfy many kinds of people. " Industry"- "interactive"Break the closed area and to Repair integrity of city life interface, all of these made The region with an open attitude interactive with the public.

# 上海世博会江南造船厂改造设计

标题标题标题标题标题标题标题标题标题标题标题标题标题标题标题标题
TitleTitleTitleTitleTitleTitleTitleTitleTitleTitleTitleTitleTitleTitleTitleTitleTitleTitleT

### 个性
#### identity
保留个性十足的工业厂房，充分利用滨水地带的自然优势，构建一个个性十足的区域.

### 延伸
#### extension
新区域将不会被孤立，并与邻近地区相互支持与补充，在城市网络引入的同时将商业，文化，旅游接通至此区域.

### 灵活的片断
#### flexibility fragment
区域原有的大空间为偶然事件提供了无限可能。一系列充满吸引力的功能空间，以混合的方式交叠在一起。

## 景观
### landscape
在滨江与城市之间的绿色景观带为市民提供更多的休闲的空间。绿植与工业遗产两者相得益彰，尽显野趣。

## 娱乐
### entertainment
重新激活原工业厂区里面的设备，改造成休闲或运动所用的设施.

## 生动
### live
从生活带动博览出发，引入人们生活中需要的功能，满足了不同人群的不同需要，在世博会之后仍然保持它的活力.

## 互动
### interactive
打破封闭的区域,缝补完整的城市生活界面,使区域以一个开放接纳的态度与市民互动。

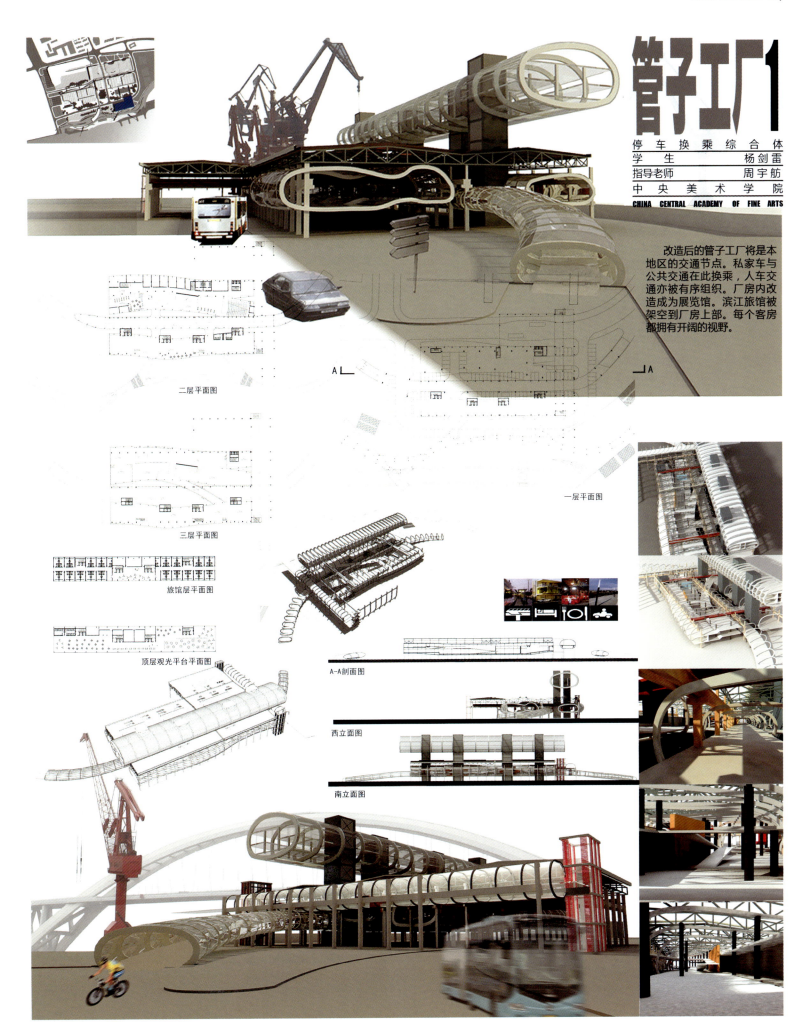

生活让城市更美好

# 管子工厂2

叠

| 学　　　生 | 曹　卿 |
| 指导老师 | 周宇舫 |

中　央　美　术　学　院
CHINA CENTRAL ACADEMY OF FINE ARTS

展览空间

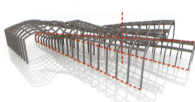

世博会是各种事件，人，叠合在一起的盛会

**改造策略：** 在旧结构的基础上叠加新结构，两种结构相互支撑，融合。并且增加了空间的丰富性

沿江立面图

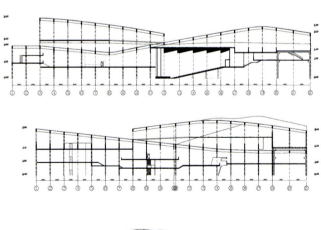

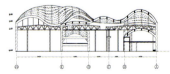

# 管子工厂 3

### 中国现代工业设计展览馆

学　　生　　　张玉婷
指导老师　　　刘彤昊
中央美术学院
CHINA CENTRAL ACADEMY OF FINE ARTS

Better Life Better City

管子工厂周边环境分析图

原有厂房（管子工厂） → 拆除部分厂房，保留具有特征的部分 ＋ 加入新的空间组织方式和互动式流线 ＝ 新旧建筑之拼贴，复合的空间关系和空间意义

总平面图

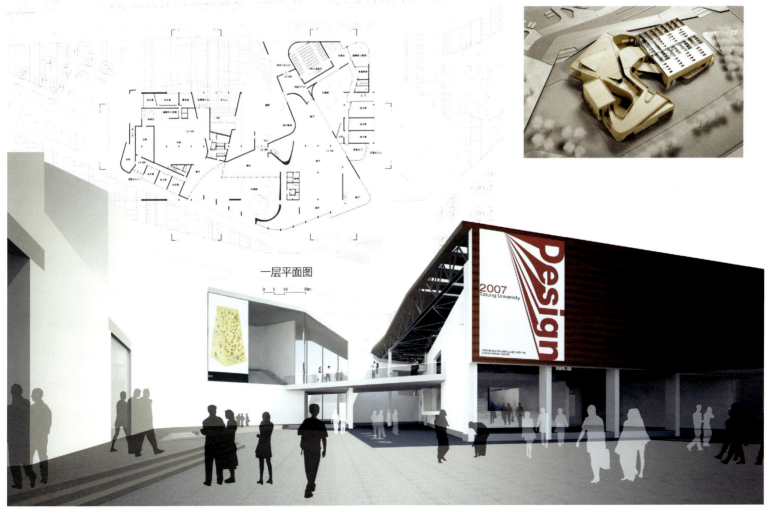

一层平面图

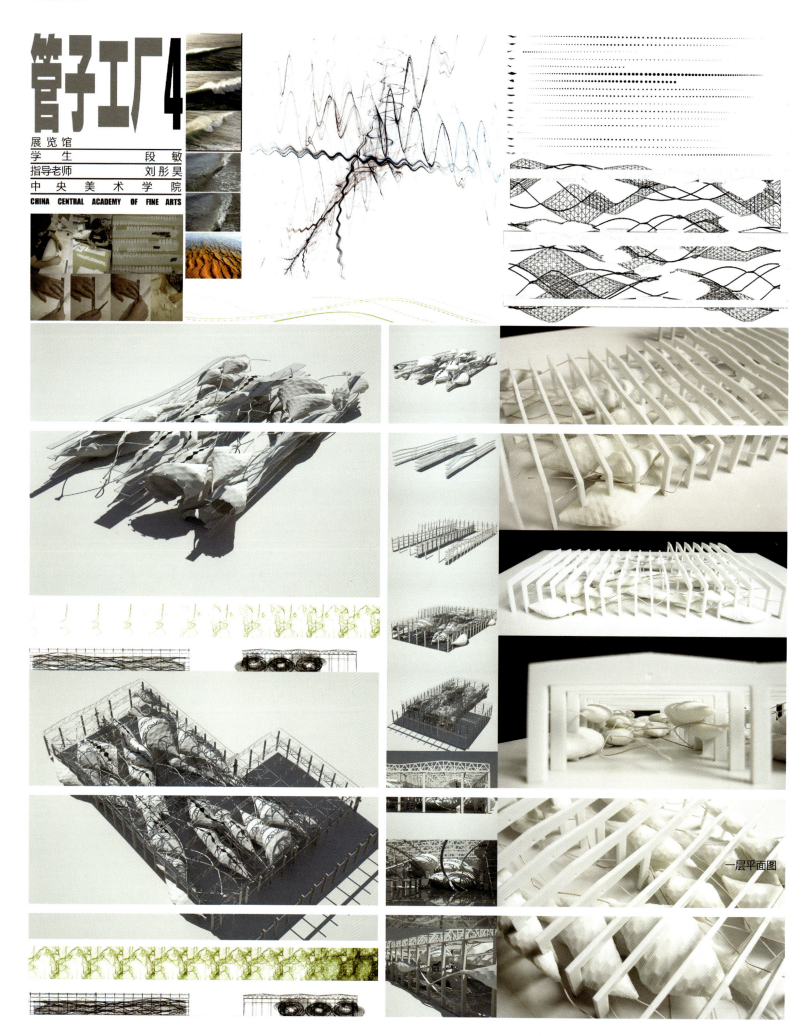

Better Life Better City

# 管子工厂 5

航模活动中心综合体
学　　生　　　　　袁　野
指导老师　　　　　刘彤昊
中　央　美　术　学　院
CHINA CENTRAL ACADEMY OF FINE ARTS

A　一层平面图

二层平面图

A-A剖面图

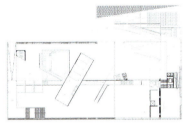

三层平面图

B-B剖面图

四层平面图

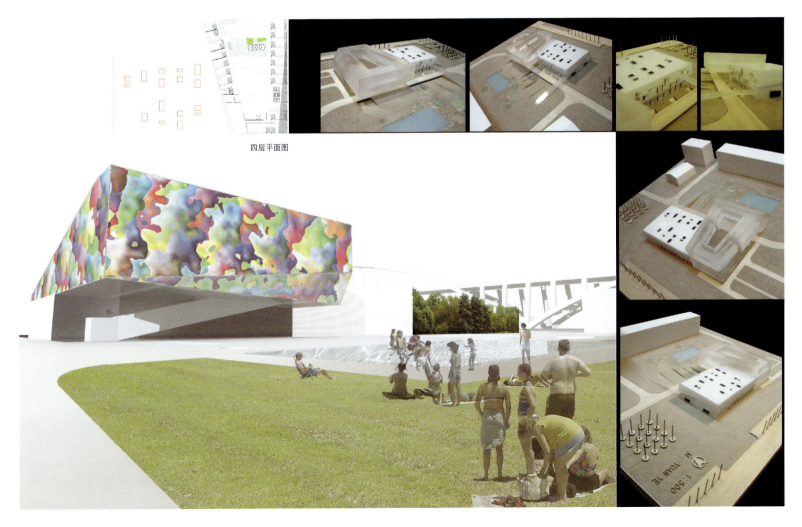

185

生活让城市更美好

# 管子工厂6

SOHO

学　生　　　马　超
指导老师　　吴晓敏
中央美术学院
CHINA CENTRAL ACADEMY OF FINE ARTS

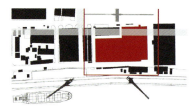
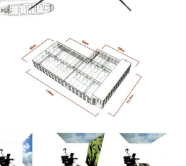
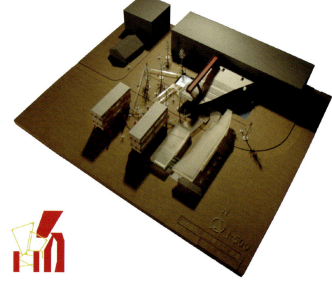

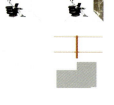         

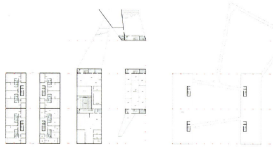
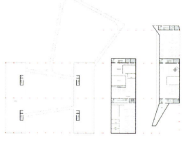
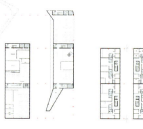
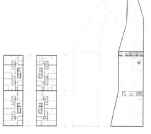
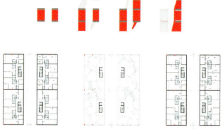

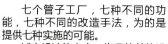

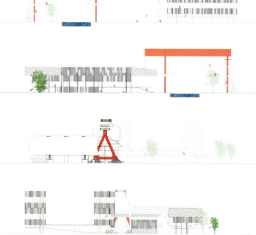

　　七个管子工厂，七种不同的功能，七种不同的改造手法，为的是提供七种实施的可能。

　　城市设计的内容，为我的单体设计提供的重要的依据，在此，我想要做的东西，被冠以了城市设计当中被重点提及的"生活"。于是，原先想要做的办公区变成了办公生活区。幸运的是，现实当中已经存在这种建筑——SOHO办公楼。

　　我依据现实当中的SOHO办公楼进行设计的同时，做了一点小小的改变，就是通过各种手段来增加生活的气息以对应"LIFE"。

　　为了增大的绿化面积，我选择了底层架空；为了提供更多的室外活动场地，我添加了篮球场和空中网球场；为了给办公的人们提供乐趣，我架起了空中跑道以供晨练；为了吸引不辞辛苦在家办公的人们出来活动，我做了四个屋顶乐园；为了体现这里的与众不同（工业遗址），我又添加了观江平台……

　　所有的这些，只是我希望在这里能够看到这样的一番景象：一个工业博物馆的屋顶会看到一个网球场！一个青少年活动中心里面的航模馆旁是人头攒动的游泳池！而在你上班办公的大楼里，你出门接杯咖啡居然看到从玻璃幕墙外跑过的邻居——他正在沿着我精心设计的空中跑道进行晨练！

# 管子工厂

**管子工厂改造 ——公共大食堂**
学　　生　　　　苏　迪
指导老师　　　　王　铁
中央美术学院

改造的绩建筑单体是位于整个设计地块东南端的管子工场，世博期间此区域的主要入口位于西北角；管子工场处于企业馆区和亲水景观区的转折处，因此在特定的时间段里工场区域范围内会聚集大量需要就餐和休息的人群，这样管子工场便充当了一个供人们休息、就餐、娱乐的"公共大食堂"。世博后作为工业博物馆群一员的管子工场还需具备临时展览的功能，这时当"大食堂"地面上的桌椅都被工场中原有保留下来的移动起重机吊起时"食堂临时展览馆"便展现在人们的眼前。

C-C 剖面（临时展览空间）

C-C 剖面（公共大食堂）

A-A 剖面（交通）

B-B 剖面（厨房）

# 同济大学
## Tongji University

吴志强
同济大学建筑与城市规划学院教授
同济大学建筑与城市规划学院院长
Wu Zhiqiang
Professor
Dean
College of Architecture and Urban Planning

吴长福
同济大学建筑与城市规划学院教授
同济大学建筑与城市规划学院副院长
Wu Changfu
Professor
Associate Dean,
College of Architecture and Urban Planning

黄一如
同济大学建筑与城市规划学院教授
建筑系副主任
Huang Yiru
Professor
Associate Chairman,
College of Architecture and Urban Planning

袁烽
同济大学建筑学院副教授
Yuan Feng
Associate Professor
College of Architecture and Urban Planning

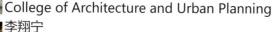

李翔宁
同济大学建筑学院教授
Li Xiangning
Associate Professor
College of Architecture and Urban Planning

联合毕业设计教学对于建筑学专业的学生有着重要的意义。它提供了一个良好平台，展现了各个学校不同的教学方法与实践。教师们可以从中看到各校的教学特点，查有余而补不足；同学们在设计思路，表达方式，方案讲解等多方面都得到了启发和提高。这次联合设计得到了兄弟学校东南大学、浙江大学的协助，增强了广泛的建筑教学与文化的广泛交流。

今年的教学成果与过程特别的丰富。上海、南京、浙江三地的三次汇合，除了设计的内容外，三校都在参观、评图过程中组织了当地的参观活动和讲座，既增添了不少欢声笑语，又为整个过程的增添了学术意义。对比中期与最终汇报的成果，我们可以发现每次碰撞带来的强烈震撼。期待下一次的联合设计教学的活动更加的丰富与精彩！

Multi-School Union graduation design has great significance to architectural education. It provides an excellent platform for displaying various ways of teaching and practicing in different schools. Not only teachers from different schools can learn from each other's strong points to offset their own weakness in teaching, but also students may get enlightened and make much progress in all respects such as ideation, approaches of expressing, and the ability of presentation.

The collective design this year got strong supports from the brother schools including Southeast University, Zhejiang University and so on, which brought profound discussion on teaching methods and comprehensive cultural communication as well. This year, the three joint presentations and appraisals in Shanghai, Nanjing, and Hangzhou were well organized along with lectures, forums and sightseeing during the whole process, which added so much fun to the academic atmosphere. Comparing the final outcomes with the interim designs, it is amazing to find that there had been so much sparkle of mind in collision of culture and concepts. The collective design offered quite a fulfilling experience and rich rewarding to everyone joined in it.

Let's looking forward to the next collective design, and hope it to be even more successful and wonderful with greater achievements.

# 清华大学
## Tsinghua University

许懋彦
清华大学建筑学院教授、建筑系主任
Xu Maoyan
Professor
Chairman of Department of Architecture
School of Architecture Tsinghua University

刘念雄
清华大学建筑学院副教授
Liu Nianxiong
Associate Professor
School of Architecture, Tsinghua University

张利
清华大学建筑学院教授
Zhang Li
Professor
School of Architecture Tsinghua University

卢向东
清华大学建筑学院副教授
Lu Xiangdong
Associate Professor
School of Architecture, Tsinghua University

作为第一次多校联合毕业设计教学活动的组织者之一，每每会对本次教学活动与前次活动有所比较，颇感欣慰的是同济大学教学团队借天时地利为本次活动贡献了一个更适合毕业设计教学的综合性课题。

印象深刻的是同济学生团队在最终交流汇报中亮出的"包袱"。非常欣赏他们在课题教学中将城市设计与分项单体设计分阶段，各组之间构成上下家相互交换作业的教学方式，看似轻松之举，着实大大强化了过程中学生之间对各种设计理念的理解与交流。

想来多校联合毕业设计教学中最大的希冀是各校师生间的交流，试想同济式的上下家"换岗"方式在下次的活动中能拓展到校际之间，岂不是一个教学过程中大范围的"强迫"交流，是否也会让人有更多的期待。

强调差异应当是多校联合教学的特色所在，这不仅是各校教学风格之间的差异，我想更应在具体教学要求、设计理念及过程方法上，可与更具体现实操作意义的城市或建筑设计要求略远一点，诸如离设置"导则"更远一点，而与畅想就能更近一点。想到清华团队的"市·博"方案，从"翻地覆天"入手，可否是一个较好的实践例证。

第一次守着北京"主场"没有太强的感觉，想到此次活动中三次南下的辛劳，更想感谢同济、东南、浙大三校师生的周到。感谢也正在化为一种期待，期待下一次成功的交流。

As one of organizers of the Multi-School Graduate Design Course, I would be very gratified to say that Tongji Architecture Department has provided us an all-around project which is more suitable to our graduate design course, compared with the project last year.

I was very impressed by the teamwork of Tongji students. They divided their work into two parts, urban design and architecture design. And they exchange their works with each other for more effective communication. Although this is not very heavy work, they can reach their goals more easily, and enforce the comprehension and communication of many architecture concepts.

I believe the key point of our Multi-School Graduate Design Course is the communication among students and teachers. And I also believe that if most the teams have done their work by Tongji's style, a more effective working style, a deeper communication would be appear through the process of our course, which surely make everyone be more excited with their works.

Emphasizing the differences is one of the key features of the course, which is not only the differences of different schools, but also the differences between our course and the practical architecture or urban projects. For instance, according to the Tsinghua team's project 'Shi-Bo', I think if there are fewer roles, there would be more imaginative works.

Finally, I appreciate all the hard work we have done, and I also appreciate the kindness of all the students and teachers of Tongji, Dongnan, and Zhejiang Universities. And I am expecting the next success.

# 东南大学
## Southeast University

王建国
东南大学建筑学院教授
建筑系主任
Wang Jianuo
Professor
Chairman
School of Architecture  Southeast University

仲德崑
东南大学建筑学院教授
Zhong Dekun
Professor
School of Architecture  Southeast University

龚恺
东南大学建筑学院教授
Gong Kai
Professor
School of Architecture  Southeast University

这次的七校联合毕业设计，使大家有机会接触到一个世博会的大项目，这样的课题可以说是世界性的。同时学生有机会在三所不同的学校进行开题、中期和期末答辩，这样的走动使学生的接触面广了，从中也接受到一些新鲜的东西。

在教学过程的两次答辩中，我听到和看到了各校学生的不同创意，每个工作组设计的推进都给我们的教学以有意义的启迪，印象尤其深刻的是同济大学的教学，他们前后两阶段的教学衔接非常有效，教师和学生、学生和学生之间的互动使得最后的成果给人耳目一新的感觉。

毕业设计结束后，我私下听取了一些本校学生的意见，大家印象最深的竟然是毕业组内实行的"一周三改"制度，因为这次东南大学有王建国、仲德崑和我三位教师参加，我们三人每隔一天就交叉下组改一次图。刚一开始，有些同学不适应，但到结束时，几乎所有的学生都认同了这样的教学密度，有三位参加完毕业设计就出国留学的学生出去后不约而同地给我来信谈到这一点，他们认为在毕业组中的训练使他们很好地适应了国外的紧张学习生活。这说明我们平时的建筑设计教学强度还不够。如何解决"一周双改"中学生设计不连续的问题是值得我们深入思考的。

最后，感谢同济大学黄一如等老师的有效工作和组织，使得本次教学能取得这样的成果。

We had a good opportunity of accessing the great project of the World Expo during the course of joint graduation project, which is attended by 7 schools. Such topic is deemed to universal. Meanwhile, students were able to precede opening, medium-term, and terminal of replies in three different schools. This kind of visit had given help in expanding the ranger of the students' view, and having fresh eyes to the outside.

During two times of replies of the teaching process, I had received lot creativity from students. Our teaching was inspired with development of each design group, especially in the teaching process of Tongji University. They had a good linkage of two teaching stages. Both the interaction of the students themselves and that of the students and the teachers made a good effort to the final result, and gave us a fresh feeling.

After the graduation project, personally I received some opinions from our own school. The most impression one was about "three times checking in one week "system (Three teachers of school of architecture SEU: Wang Jianguo, Zhong Dekun and I had attended the teaching courses, each one would check the design of different groups every two days). In the just beginning, some students were not suited, but at the end almost all the students had identified with such studying density. After the graduation project, there were three students who were pursuit for further studying abroad coincidentally lettered to me. They considered that the training in the graduation project developed a good ability for them to adapt the stress life pace in foreign countries. It seems that we can strengthen the intensity in our teaching course of architecture design. So, it is very necessary to think about how to solve the problem of discontinuity in the design course of "two times checking in one week "system.

Finally, I sincerely appreciate teacher Wang Yiru of Tongji University etc., with whose effective works and organizations we achieved such successful results.

# 天津大学
## Tianjin University

张玉坤
天津大学建筑学院教授
Zhang Yukun
Professor
School of Architecture  Tianjin University

青木信夫
天津大学建筑学院教授
AOKI Nobuo
Professor
School of Architecture  Tianjin University

徐苏斌
天津大学建筑学院教授
XU Subin
Professor
College of Architecture and Urban Planning

谭立峰
天津大学建筑学院博士
Tan Lifeng
Ph.D
College of Architecture and Urban Planning

多校联合毕业设计作为各校教学交流的平台已经步入第二个年头。我们作为参加此次毕业设计的指导教师，有几多感慨，也有几多收获。纵观此次联合设计，可以说喜悦伴随着缺憾，缺憾孕育着希望。

喜悦——在此次联合设计中，我们看到了各校学子颇有创意的方案。其中，多数方案都有比较严密的逻辑性和思想性，这反映了学生扎实的理论基础和锐意创新的意识。另一方面，各校师生利用这一平台，相互交流设计教学经验，从而取长补短，取得了一般毕业设计难以取得的收获，这是可喜可贺之事。

缺憾——此次联合设计也显露出了需要改进的问题。大家共同探讨、相互学习的机会还不是很充分，交流的形式还比较单一；各校毕业设计成果要求的不同也是此次暴露的问题，致使成果深度难以保持一致……

希望——我们毕竟尝试了创新，发现了问题，积累了经验，在以往的基础上又前进了一步。取得的成果尽管还显得几分青涩，但这是一笔教学创新的财富。我们期待能够进一步加强交流，改变单一的汇报模式，在下一次的联合设计中取得更大的进步。

总之，我们在用坚毅的脚步踏出一条毕业设计的探索之路，为中国的建筑教育做出自己的一份贡献。

As a teaching exchange platform, multi-school joint graduation design has entered its second year. Being instructors of the graduation design ,we have some mixed feelings, and obtained fruits as well. Throughout the joint design, it can be said that the enjoyment accompanied with imperfect, whereas the imperfect will beget hope in the future.

Enjoyment— In the joint design, we have seen quite innovative designs done by the students. Most of the projects are relatively more logical and thoughtful, reflecting the students' a deep theoretical foundation and a keen sense of innovation. On the other hand, school teachers and students exchanged experiences in teaching and learning on the platform, thus learning from other's strong points to offset one's weakness, and achieved much more than an ordinary graduate design .Congratulations!

Imperfect— The joint design also revealed some problems need to be improved . Opportunities are not full enough for the students to discuss together and to learn from each other; the form of exchange is still relatively single; and because of different requirements of the graduation design in different schools, the outcome depth of the designs exposed inconsistencies, leading difficulties to maintain consistent results ……

Hope — We, after all, have gone through an innovative process. We have found some problems and accumulated experiences during the procedure, getting a step forward on the basis of the past . Although the results appeared seem not very perfect, this is a wealth of teaching innovation. We are looking forward to Strengthening further exchanges and will try our best to achieve greater progress in the next joint design.

In short, with our great determination and efforts, we are searching a way for the graduate design, and lending a hand to contribute the architectural education of China.

# 重庆大学
## Chongqing University

卢峰
重庆大学建筑城规学院建筑系主任
教授
Lu feng
Professor
Dean of Architecture Faculty

顾红男
重庆大学建筑城规学院副教授
Gu Hongnan
Associate Professor
School of Architecture  Tianjin University

七校联合毕业设计在紧张有序的氛围中圆满结束了。回头想来，天马行空的巧妙构思、不同设计理念的激烈碰撞，各校同学在讲台上的出色演讲，都给初次参与这一活动的重大师生留下了深刻的印象。在此要特别感谢同济大学、东南大学、浙江大学的老师和同学们，正是他们辛勤的付出和有效的组织，才使我们的每一次聚会都充满了激情和欢声笑语。

近几年来，对应于与国外院校的频繁教学合作，国内院校之间的交流渠道相对较少，联合毕业设计的教学模式，为国内不同地域的建筑院校之间相互了解与观摩，提供了一个直接而有效的平台。此次七校联合毕业设计的选题和选址都具有很强的现实意义和针对性，使我们的同学在世博会的宏大叙事下，获得了一个综合运用所学知识解决实际问题的切入点，在设计过程展开、阶段成果汇报、调动同学们的创造积极性等方面，均起到了预期的教学效果。

今天的收获使我们对新的课题充满了期待，期待每一次联合设计都有一个引人入胜的选题，引导我们全身心地面对新的挑战；期待联合设计的教学交流能像一面镜子，帮助我们更深入地了解自己的特色与不足，从而在这个同一化的时代里，保持自己与生俱来的特质，并为我们的教学改革增添新的思路和方法。

Today's experiences leave us expectations for the new topic. We expect that each united design will have an enchanting topic that guides us to face the new challenges with all our hearts. We expect that the teaching exchange of united design will help us have a more thorough awareness of our specialties and shortcomings, so that we are able to maintain our inherent qualities in this oneness age. The united design will add new thinking and methods to our teaching reform.

The united graduation design launched among seven schools is successfully concluded in an intense but orderly atmosphere. Thinking back, the unconstrained creative ideas, the fierce impacting of different design concepts and the wonderful speeches of the students have imprinted deep impressions to the teachers and students of Chongqing University, who participate in this activity for the first time. Here we would like to extend our special thanks to the teachers and students from Tongji University, Southeast University and Zhejiang University. Thanks to their hard efforts and effective organization, each of our meetings is full of enthusiasm and laughter.

In contract to frequent exchange with foreign institutes in recent years, the exchange between domestic institutes is relatively less. The teaching mode of united graduation design provides a direct and effective platform for the domestic architectural colleges in different regions to know and learn from each other. The topic and address of the united graduation design are selected with highly realistic meaning and pertinence, so that we are offered a breakthrough point to use what we have learnt to solve the realistic problem under such a grand background – the World Expo. United design has achieved the expected teaching effect in all aspects from development of design process to reporting of by-stage result and inspiration of the students' creativity.

# 浙江大学
## Zhejiang University

王竹
教授
浙江大学建筑工程学院副院长
建筑学系系主任
Wang Zhu,
Professor
Associate Dean of College of Civil Engineering and Architecture
Head of the Department of Architecture, Zhejiang University

朱宇恒
浙江大学建筑学系副教授
建筑学系主任助理
Zhu Yuheng
Associate Professor
Assistant of the head of the Department of Architecture, Zhejiang University

建筑学专业联合毕业设计是一个较新的教学环节，对学生们开阔眼界、拓展思路、寻找差距都有积极的意义。我校是第一次参加联合毕业设计，建筑系毕业生积极踊跃报名，最终选择了六分之一的学生参与这个题目进行毕业设计选题，表现出极大的热情。本次七校联合毕业设计在去年第一次举办的基础上又有了新的举措，将开题、中期交流和最终成果答辩分别放在三个不同地点的学校举行，使活动更具广泛的交流性。

在此次活动的各个阶段，我们都看到了学生们活跃而开放的思维。通过几乎涵盖了城市设计所有元素的成果，此次联合毕业设计可谓精彩纷呈，体现了不同学校的教育特色和设计风格。虽然在如何最终体现毕业设计深度上还有待探索，但联合毕业设计对打破各校之间的信息交流盲区、促进国内主要建筑院系师生的相互了解、交流和提高建筑学教学水平都有着极大的意义。

Joint graduate design in the major of architecture is a brand new move in teaching for the students in our university. It plays an active part in broadening their visions, expanding their ideas and finding the gap between themselves. Though it is the first time for us to join such a graduate design, students in architectural major apply to take part in it with great passion and only one sixth of them are chosen to be the lucky dogs at last. Compared to what it was last year when it was initiated, the 7-university-joint-graduate-design decides to hold the three stages of graduate design —opening, middle-term exchange and oral defense—in 3 separate universities, which makes it possible for the 7 universities to communicate with each other more extensively.

This activity completely shows our students' open mind and smart ideas throughout the three stages and bears fruitful results in all elements of urban design. Different universities have shown their educational characteristics and design styles. Though we still have a long way to go in the field of how to complete the graduate design to a very great extent, the move of the joint graduate design has great significance in breaking the blind spots where exchange is not fully carried out among universities, in improving the teaching levels of major domestic faculty in different architectural universities, and in promoting mutual understanding and exchange between students and teachers of major architectural departments around China.

# 中央美术学院
## China Central Academy Of Fine Arts

吕品晶
教授
中央美术学院建筑学院院长
Lu Pinjing
Professor
Dean of the School of Architecture of CAFA

刘彤昊
中央美术学院建筑学院讲师
Liu Tonghao
Taecher
School of Architecture of CAFA

周宇舫
中央美术学院建筑学院副教授
Zhou Yufan
Associate Professor
School of Architecture of CAFA

王铁
教授
中央美术学院建筑学院副院长
Professor
Associate Dean of the School of Architecture of CAFA

傅祎
副教授
中央美术学院建筑学院副院长
FuYi
Associate Professor
Associate Dean
School of Architecture of CAFA

吴晓敏
中央美术学院建筑学院副教授
Xiaomin Wu
Associate Professor
School of Architecture of CAFA

中期到南京,火车过天津,随手拍了张照片,后来整理照片时才发现是天大参加七校联合毕业设计的学生和老师正在登上同一次火车,在相邻的车厢。终期到杭州,好难买到火车票,坐到杭州,后来知道清华的同学没赶上同一列火车,又被大雨阻在沪杭高速上……或许,没有别的人会在意这些小事件,就像我们不知道其他的乘客的目的地一样,但是,七校联合毕业设计,把我们几个学校的师生联系了起来,特别是参与其中的同学,都很在意有过的经历,视为人生的重要一步。

在美院老师们的眼里,在意的更是各个学校毕业设计教学方法的异同,学习和借鉴,以及把自己的教学成果呈现在中国最优秀的建筑院校的师生们面前,展示自己,听取意见和建议。重在参与的精神不仅仅是做出好的作品,也要贡献出美院师生的才智和艺术感觉,为这个活动增加几分来自艺术院校的色彩。对于我们教师而言,是回了家的感觉,因为美院的大多数老师都毕业于参与的几个学校,新朋老友、师长学弟,找到了归属感,我们很开心。

终期汇报时,美院的苏迪同学在讲解自己的万人大食堂提案时,引起了会场严肃气氛的改变,一时间,大家都很高兴。或许这就是所谓的"美院色彩",在认真、艰苦地找寻感觉之后,将一切调和成悦目的表象。

We went to Nanjing for the Midterm review by train. When our train stopped by Tianjing, we took some pictures at the station. When we examine these pictures later, It happened that we caught the students and teachers from Tianjin University who were boarding the same train at the adjacent compartment. The final review was taken place in Hangzhou. While it was difficult to buy train tickets and finally we had to take the seating compartment of the train, which took one night. When we arrived in Hangzhou, we heard that the students of Tsinghua University did not catch the same train, and as well stopped by heavy rain at the Shanghai-Hangzhou Expressway…Perhaps, no one would pay attention to those trivial things, just like we would never know the destinations of other passengers on the train. However, the Seven-School-Joint-Graduation-Project, connected our teachers and students with each other from seven universities. Especially the students involved paid enough attention to the whole process, and considered it an important milestone of their lives.

From our points of views, we concerned more about the similarities and differences between the various schools' teaching methods and tried to learn from others. At the same time, we were to show our teaching results to those from outstanding institutions, to obtain their comments and suggestions. Our purpose were not only participating through good works, but also contributing to this event the talent and artistic impression of teachers and students from CAFA, as well as increasing a fraction of color to this event by the characteristics of art institutions. For our teachers, the feeling of coming back home schools was additional because the majority of teachers in the school of architecture of CAFA have graduated from other participated schools. We were very happy of finding the sense of belongings.

During the Final presentations, a student from CAFA named Su Di, gave a speech on his proposal of a food court for ten thousand people, which broke the serious atmosphere of the conferences room, and cheered up most of people in that room. Perhaps this was so-called "color of art academies" — after a serious and painful search, reconcile everything to a pleasing appearance.

图书在版编目(CIP)数据

走进EXPO2010：2008七校联合毕业设计作品/黄一如等编．—北京：中国建筑工业出版社，2008
 ISBN 978-7-112-08178-3

Ⅰ.走… Ⅱ.黄… Ⅲ.建筑设计-作品集-中国-现代
Ⅳ.TU206

中国版本图书馆CIP数据核字（2008）第149003号

　　本书记录了同济大学、清华大学、东南大学、天津大学、重庆大学、浙江大学、中央美术学院七所学校建筑学专业本科毕业设计"城市建筑的更新与再生——2010年上海世博会中国现代工业博物馆群设计"的教学内容。该课题包括城市设计与建筑设计两大部分，学生从整体区域研究出发，通过现场调研、问题分析、项目策划、提出城市设计方案，并有选择地完成建筑单体方案的设计。通过规划统一设计，对江南造船厂区域的工业遗产在2010年世博会的定位和发展作一个探讨研究。

　　本书内容有教学活动过程的记录和介绍，设计的任务书，7所院校16份作品以及各校指导老师的课程感言。可供国内外建筑院系建筑教学交流、参考，也可供相关专业师生借鉴与学习。

责任编辑：徐纺
责任校对：李品一

**走进EXPO2010：2008七校联合毕业设计作品**

黄一如　许懋彦　龚恺　徐苏斌　卢峰　王竹　吕品晶编
\*
中国建筑工业出版社出版、发行（北京西郊百万庄）
各地新华书店、建筑书店经销
上海界龙艺术印刷有限公司制版、印刷
\*
开本：965×1270毫米　1/12　印张：17　字数：520千字
2008年11月第一版　2008年11月第一次印刷
定价：**108.00**元
ISBN 978-7-112-08178-3
　　　　（14132）

**版权所有　翻印必究**
如有印装质量问题，可寄本社退换
（邮政编码　100037）